Scrapie and Mad Cow Disease

The Smallest and Most Lethal Living Thing

G. D. Hunter, D.Sc.

VANTAGE PRESS
New York

FIRST EDITION

Published by Vantage Press, Inc.
516 West 34th Street, New York, New York 10001

Manufactured in the United States of America
ISBN:0-533-10230-8

Library of Congress Catalog Card No.: 92-93299

0 9 8 7 6 5 4 3 2 1

To my wife, Joy

Contents

Acknowledgments

It is a pleasure to be able formally to thank so many people who helped in the preparation of this book. First and foremost comes my wife, Joy, who read every version of the text and made many valid and useful criticisms. Without her constant encouragement, I would never have completed the work.

Mrs. Kathy Le Brun has somehow managed to decipher what passed for my handwriting and has been an ideal secretary and typist. Mr. David Craig, the former secretary of IRAD (now called the Institute of Animal Health, IAH), has helped in many ways, not least in reading and commenting on the final version of the text. I must thank Dr. Michael Rutter for permitting me to use the library at IRAD after my retirement.

My grateful thanks go to all who have read and without exception made helpful comments on the text: Ms. Carol Walker (who read two versions), Dr. David Baird, Dr. Paul Britton, Mr. Michael Clarke, Dr. Alistair Lax, Dr. Jane Manning, Mr. Geoff Millson, Dr. Kevin Page, and the late Prof. Jack Payne.

I have incorporated so many suggestions from other people that the book is as much theirs as mine. I also found invaluable James Parry's book on *Scrapie in the Sheep* and used it extensively as a source of information in compiling the first two chapters of this book.

Mr. Dave Hawkins was kind enough to prepare much of the illustrative material.

Finally, I must thank the many people who provided me with photographs and manuscripts, notably Prof. Stan Prusiner for letting me see copies of papers prior to their publication.

Introduction

This book tells the story of scrapie. It is a story that until recently seemed to be shrouded in a conspiracy of silence as far as the general public was concerned. Mad cow disease has changed all that, and everyone can now share the great interest and excitement that scientists have derived from the subject for several decades. The excitement is generated in several ways: in the first place, we have in scrapie perhaps the most mysterious of disease agents. Even though the brain of a small animal infected with the disease can contain no less than a trillion infectious units of the disease, we still do not know what it is, an almost unbelievable circumstance in these days of advanced molecular biology. We do know that it is smaller in size than any known virus, and there has been much speculation that it may somehow reproduce without itself possessing either of the standard genetic materials, the nucleic acids DNA or RNA. We understand, however, a great deal about what scrapie and related disease agents do. They are found throughout the body of the animals they infest, but they only appear to cause significant damage within the brain and the rest of the central nervous system. Here the key neuronal cells that control all the vital activities of life are disrupted and destroyed so that paralysis, dementia, and death are the inevitable result.

Another reason for the great fascination of scrapie is that we ourselves are attacked by closely related agents of disease. These again were little known until recent events put the spotlight on them, but Creutzfeldt-Jakob disease causes early senile dementia in the human race throughout the world, and probably at least one person in 50,000 succumbs to this deadly condition. Kuru is a related disease confined in this case to a single people, the Fore of the Highlands of New Guinea. Finally, the Gerstmann-Straussler syndrome occurs in families and seems to be similar in many ways to Creutzfeldt-Jakob disease. It is possible that several as yet unsolved human disease problems may also involve agents of the scrapie type. In particular, the great scourge of old age, Alzheimer's senile dementia, may be one such problem. Thus we all have a profound reason for wanting to know more about scrapie.

The scrapie agent is also extraordinary in the slowness of its action. It can be shown in some animals that it will multiply slowly throughout the whole lifespan, but old age and natural death intervene before disease occurs. Similarly, in humans it has been shown that more than twenty years of perfectly healthy life can be enjoyed between infection with the kuru agent and the appearance of the fatal disease. Hence the widespread use of the term "slow virus." This name is applied somewhat indiscriminately, however, to some chronic viral conditions, and hence other people prefer the term "unconventional agent."

Incubation periods of years give scientists ample time for reflection, and tension mounts during the prolonged wait for the seemingly interminable experiments to yield results. The tension has been reflected in rich displays of temperament and character by the research workers involved, and this book attempts to open a window on the scene for the benefit of the general public and to show them that scientists are first and foremost human beings just like everyone else. The human interest is not limited to the present day, and the romantic history of scrapie is entangled with many other strands of past human activity, such as the Napoleonic wars and the improvements of farm animal stock in the eighteenth and nineteenth centuries. But it is always the future that is most important, and surely we must see a solution to the scrapie problem within the next decade or so, as it bows to the pressure of ever more sophisticated new teams of research workers, spurred on by the urgency of the problems generated by mad cow disease.

I hope that this book will help the public at large to appreciate and understand the developing drama as the scrapie story unfolds.

Chapter 1
Scrapie Disease

GENERAL POINTS

Although scrapie has long been known as an untreatable and slowly fatal disease of sheep, its cause, cure, and prevention have been and still are matters of controversy. Since the discovery of the experimentally transmissible scrapie agent in the tissues of affected sheep in 1936, the view that scrapie has an infectious cause has held sway over suggestions of an inherited cause. It is now generally accepted as the representative-type disease for a group of nervous disorders of man and of domestic animals, with the following general characteristics:

1. The disease appears insidiously, most commonly in middle age or later, and previous good health is gradually replaced by a fatal disease of the nervous system.
2. Inflammation is absent, but "spongy" degeneration of the nervous system leads to a steady loss of nerve cells; remarkably, all the lesions observed on one side of the brain tend to appear simultaneously with exact bilateral symmetry on the other side of the brain as well.
3. The immunological system of the attacked victim of disease fails to respond, and there are, for instance, no circulatory antibodies active against the disease agent.

The relationship of scrapie to the other diseases will be considered more fully in chapter 3. Here we will confine ourselves to a description of the sheep disease and its manifold variations.

EARLY SCRAPIE

Premonitory signs of scrapie may appear in sheep several months before any definite illness can be recognised. In closely shepherded flocks, the possible implications of these signs are recognised by experienced shepherds, and they cull the sheep before conclusive clinical evidence of disease appears. Thus, they may keep their flocks "free from scrapie." The skilled shepherd will observe short periods of irregular changes of behaviour, which would pass unrecognised by most people. The affected sheep may stand quietly, head raised in a fixed position and staring ahead as if at some object, but the eyes remain immobile and unresponsive. While the rest of the flock graze quietly on a good grass sward, the same animal may move repeatedly and take only a few mouthfuls at each step. Similarly, it is likely to move repeatedly to new positions at a well-filled trough, while its fellow animals stay put. A similar restlessness may be shown when the remainder of the flock is quietly ruminating, and it will be the first to turn and face an intruder if the flock is disturbed.

Aggressiveness is occasionally displayed and the affected animal may become difficult to herd and drive and, as it becomes agitated, even charge the sheepdog or gate. Scrapie is sometimes detected very early by the shearer, where the small "jittery" movements displayed by the restless animals can cause him problems.

Eventually, early clinical disease becomes established, but here again it is frequently inconspicuous, with the fleece of affected animals showing perhaps a whitish tip to the staple over the back: it may lose springiness and "character," becoming more open and with a harsh "handle." These changes probably reflect small alterations in the nutritional biochemistry of the skin. The casual observer will not easily detect such scrapie animals in a mainly normal flock.

Although there is initially no obvious loss of bodily condition, the intake of water soon becomes disturbed. Animals visit the water trough frequently but ingest only small amounts on each occasion. The tolerance of exercise is reduced, and affected animals soon tire when driven. They may even collapse in their tracks as they walk but recover after a short rest. After a week or two, the same animal may develop a clumsy gait and experience difficulty in turning sharply off its hind limbs. A little later on, the sheep often begins to rub itself against posts or any other suitable solid object. The commonest sites of rubbing include the buttocks on each side of the tail and the poll of the head between the ears. Other animals may use their hind feet to scratch areas in front of the shoulder joint or behind the elbow. But we are moving now towards the frank clinical stage of the disease, when scrapie can be easily recognised by anyone who has received a minimum of tuition in the disease.

CLINICAL SCRAPIE

Shepherds describe the animals as being "unthrifty": their hair and fleece look dull and lustreless, and they do not carry as much flesh as normal sheep with similar access to food in the same environment. Some affected animals suffer muscle wastage quite rapidly, and all are unable to walk easily more than about a quarter of a mile. Related to this lack of stamina is the onset of constitutional disorder of the limbs (ataxia), with excessive flexion of both fore and hind limbs. They tend to be placed as far apart as possible and the animal thus acquires a wide-based stance. The animals are easily upset and excited with an accentuation of any ataxia present. Other animals suffer waves of trembling or shivers that pass down the back like "someone wailing over a grave" (hence the French name for scrapie, "la tremblante"). Again the head may be held up, with the nose in the air.

Soon the signs of rubbing are clearly evident in most scrapie animals. As the wool over the rubbed areas becomes matted, the outer part of the staple is lost. Sometimes the wool fibres are shed and replaced by new short growth, which in dark-faced breeds is usually pigmented. Some writers refer to "shrugginess" when there is a general brushing of the side of the whole body, but more usually rubbing is confined to localised sites such as those mentioned above and the lateral thorax. The urge to rub is considerable, and affected sheep will stagger for quite a long way to reach a favoured post where it may indulge itself for as long as five minutes. There is a risk of other infections through all the cuts and scratches, especially around areas difficult to scratch with the feet: the saddle of the back, the ears, and the rump and tail. The sheep nibble at the haired skin below the knee and hock, and in fact any available areas are bitten and the fleece pulled out. Hence the "poodlelike" appearance this self-trimming gives to the sheep, as for instance in the animal shown in Fig. 1.

Fig. 1
Scrapie sheep showing typical "poodle" effect of rubbing on the flanks.

Characteristic of many scrapie sheep is the "nibbling" response that can be elicited by scratching the animal firmly along the back just above the tail, the pressure being sufficient to involve the underlying muscles. The animal's attention is held, and it raises its head and nose slightly and at the same time engages in nibbling movements of the lips. With lip smacking and extrusion of the tongue, the sheep seems to gain considerable satisfaction from the process. Other symptoms of the disease include the appearance of rashes and papular eruptions, sometimes developing to small hard pimples, from which hair does not regrow. The animals often retain normal appetite, with a food intake sufficient of itself to sustain them in good health. The earlier disorder of water intake is, however, accentuated, and scrapie sheep will often drink twice as much as normal animals, with a preference for salt rather than fresh water. Other physiological parameters may be normal: tendon reflexes, defecation, micturition, hearing, and sight. In some animals, however, the voice has a high-pitched timbre.

Sheep enter the terminal stages of the disease, usually, 3–4 months after the onset of the first clear clinical signs. The fatal outcome is now 2–4 weeks away in most animals. They lose weight steadily, and the muscles become small and flabby. They cannot walk more than about 50 yards at a time and are difficult to dislodge from a recumbent position. Any change in their normal routine can upset them and lead to a rapid fatal outcome, even a small change such as alteration of their pen and immediate companions. Although the ataxia is accentuated and they fall regularly, muscular power is not necessarily lost and one affected ram was reported by H. B. (James) Parry to have cleared a 4-foot-6-inch-high oak farm gate only 4 days before its death from scrapie. A clumsy kind of gallop is often used with two hind limbs moving together, reminiscent of the action of a springhorn antelope; but with only one leg on the ground, balance is lost. Normal posture is still adopted at rest.

Later still, the sheep lose even the clumsy ability to regain their feet, but if housed and fed can survive for a week or more lying on their sides. Some lose the ability to eat altogether and of course animals in the field die rapidly of exposure. Although most animals die 4–6 months after the first onset of clinical signs, maternal pressures can hold scrapie at bay. A pregnant ewe with severe symptoms will bear her lambs and nurse them to independence before succumbing to the final stages of the disease. But there are wide variations, and occasionally mothering behaviour is lost in a scrapie ewe after giving birth to her lambs. Males are fully fertile and active well into the clinical stages of the disease, and indeed their libido is maintained fully to the end; in fact they tend to be more active at mating time than corresponding normal rams. Control of body temperature and heart rate are less good than in normal sheep.

Many other symptoms are shown by some animals or groups of animals, and for instance it was at one time thought that there was a relationship to human muscular dystrophy. But the analogous clinical signs were not found to be regularly observed. Other scrapie animals exhibit defective vision, hind limb paralysis, epilepsy, and loss of control of the nervous system. The clinical diagnosis of scrapie is usually clear to the expert; but there is always a grey borderline area in any disease, and in the case of scrapie the disease can only be confirmed by examining the brain after death under the microscope. Even then, there are rare cases of uncertain diagnosis.

ONSET, COURSE, AND CONFIRMATION OF THE DISEASE

Although experimentally inoculated animals can contract scrapie as young as 6 months of age, the onset of the disorder in the field is rarely below the age of 2 years, while 90–95% of cases are manifest by the age of 5. Both sexes seem to be equally prone to the disease, but rams sometimes contract scrapie at a slightly younger age than ewes. The relentless course of the disease is usually 3–6 months, but a few cases of remission have been reported for periods of up to 2 years, although always with subsequent relapse and death. After the relentless slow decline,

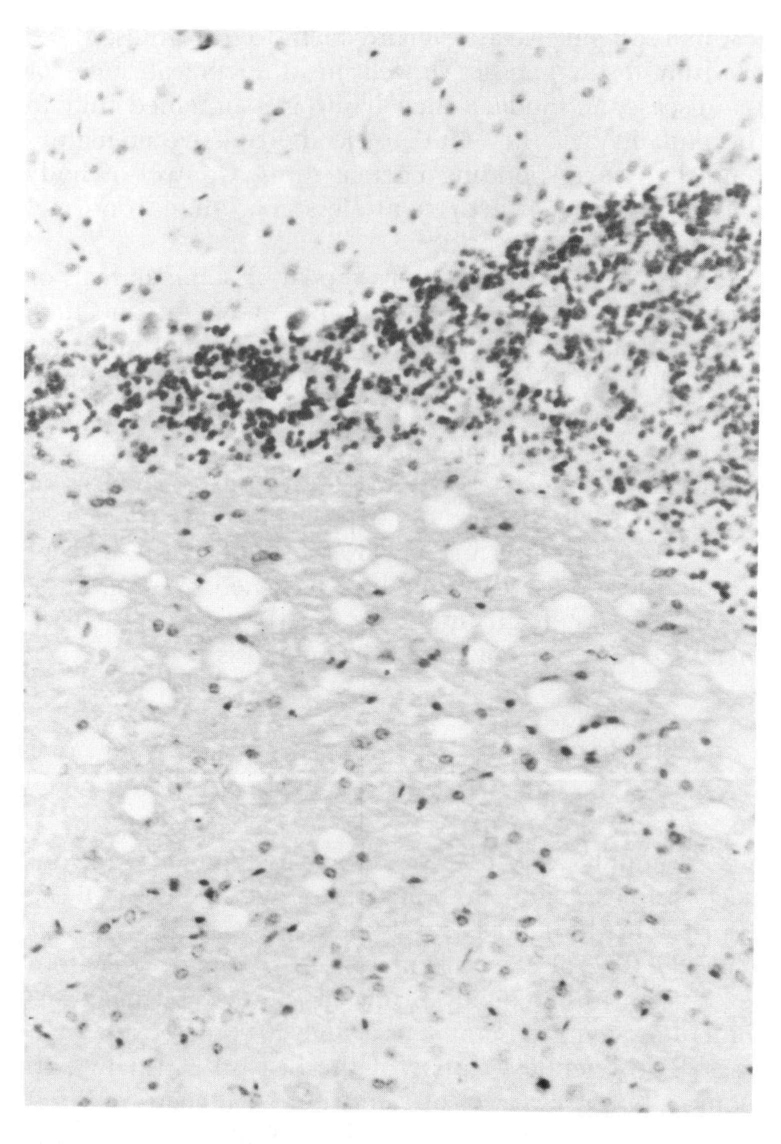

Fig. 2
Scrapie brain histology: (a) Vacuoles (in and around neurones)

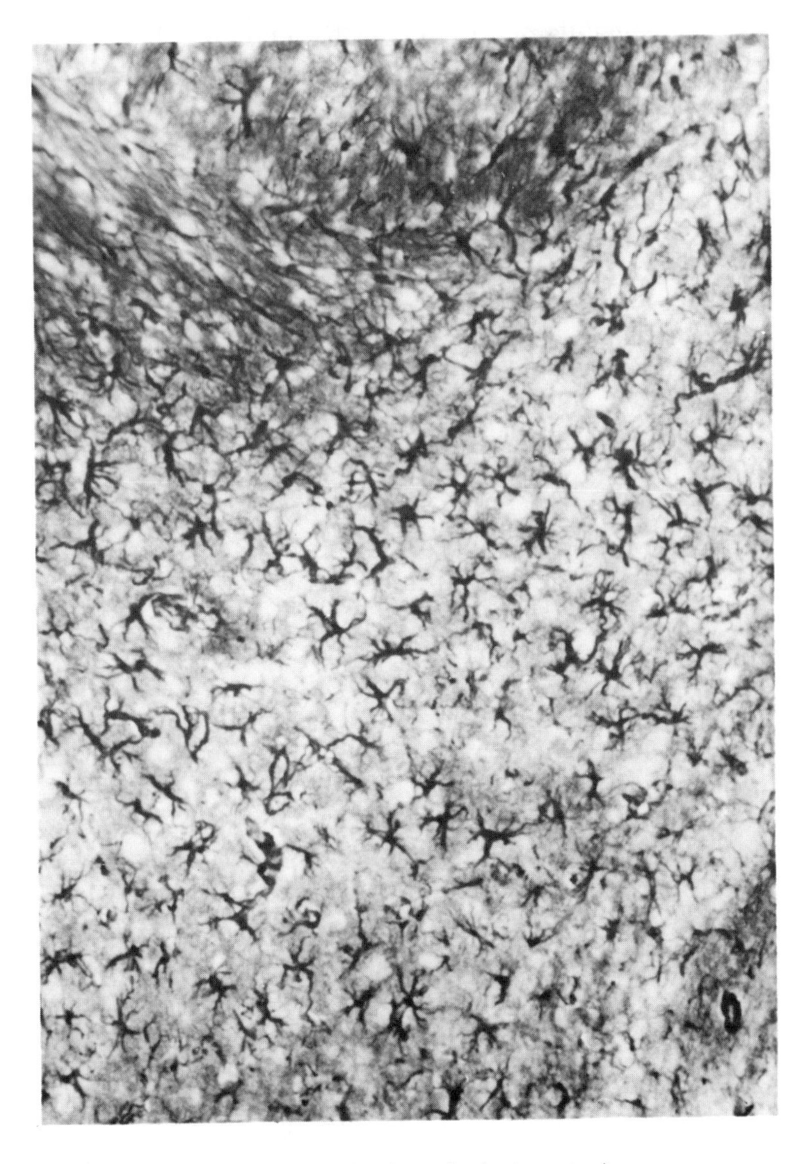

(b) Enlargement of certain brain cells (astrocytes).

appetite is maintained until near the end, and death usually occurs at night after exposure of prostrate animals to cold rain. No effective means of preventing, delaying, or curing the disease in sheep are known.

Until recently, there were no specific tests available for the detection of the scrapie agent in sheep brain. As we shall see in chapter 5, it is now possible to link the presence of the scrapie agent to a large increase in the amount of a particular protein; but the analytical and biopsy techniques involved are too complex for routine operation in the field, although efforts are currently being made to simplify the procedures. Thus the diagnosis of scrapie still has to be confirmed by microscopic examination of the brain after death.

Although the scrapie agent grows and multiplies to some extent throughout the body of the animal, it appears to damage only the brain and nervous system; and the damage there is only obvious at the end of the long incubation period when the amount of scrapie agent present has become very great. The neuronal cells seem to be particularly affected: these are the cells that are the main constituents of the nerve channels that carry the impulses that connect all our voluntary and involuntary actions (and perhaps our thought processes). They show characteristic spongy degeneration in certain areas, with vacuoles (holes) appearing. Another type of brain cell, the astrocyte, which probably participates in providing the neurone with its essential nutrients, often greatly increases in size. Microscopic examination of appropriately stained sections of the brain displays these changes in neurones and astrocytes very clearly. (See Fig. 2a and b.)

Chapter 2
A Short History of Scrapie

INTRODUCTION

Scrapie has probably been with us for as long as records have been kept of disease and pestilence in domestic animals. There are indications of references to scrapie in early reports in Latin, but it is not possible to disentangle the accounts of the various "murrains" of sheep and cattle and make precise diagnostic assignments. The first reference in modern languages appear early in the eighteenth century. An occurrence in merino sheep was reported in Spain in 1732, and about the same time it was reported under the name "rickets" in England. Later there was a devastating outbreak of scrapie in Lincolnshire, and local farmers petitioned the British Parliament to introduce regulations that restricted the movement of sheep. It is interesting that as early as 1733 it should have been appreciated that an agent transmissible between animals was likely to be involved.

As the eighteenth century progressed, so did the Age of Enlightenment, and agricultural problems were approached in a more scientific and systematic manner than hitherto. The climate also improved after the "the Little Ice Age" of the sixteenth and seventeenth centuries. Particularly affected were livestock improvement and breeding, and the observations on scrapie made by perceptive owners and shepherds of 200 years ago are immediately recognisable today by anyone who has had practical experience of the disease. However, it was probably not just better diagnosis that led to the increased recognition of scrapie in the eighteenth century, but rather, altered breeding practices are likely to have led to an increased incidence of the disease.

Efforts were being made, for instance, to improve the performance of the English Thoroughbred racehorse, based in Newmarket, and there was an innovative move away from what had previously been considered a safe low level of consanguinity with the adoption of a system of closer inbreeding. The extreme example of this was called "breeding in-and-in," i.e., the mating of father to daughter, son to mother, brother to sister, uncle to niece, etc. Such systems were adopted, for instance, by Bakewell of Dishley and Ellman of Glynde, and despite warnings from conservative breeders like Arthur Young, they did enable a potentially desirable genetic characteristic to be fixed. So long as few deleterious genetic traits were carried in the foundation stock of animals, and provided any that were present were immediately recognised and eliminated by radical culling out of affected animals, all was well. Some breeders, particularly of racehorses and cattle, were outstandingly successful during this period, but there were also disasters as when Bates's Duchess shorthorns developed an imperforate hymen with complete infertility.

Many of our major sheep breeds emerged at this time, again with mixed success. With a disease such as scrapie, manifesting in middle age, often more than halfway through an animal's reproductive life, the control of any genetic predisposition is dependent upon individual identifications and meticulous mating and lambing records. These conditions were nigh impossible to meet under eighteenth-century farming conditions. Some breeds emerged in England largely free from scrapie, and for instance the English Leicesters, Southdowns, Kent or Romneys, and Cotswolds have remained so to the present day. Other breeds were not so fortunate, and a further complicating factor was the increased importation after 1700 of the soft woolled Spanish merino sheep, many of which carried scrapie or the susceptibility to the disease into other parts of Europe. This had a particularly catastrophic effect in the Spanish merino studs of Germany and France, where the practice of breeding "in-and-in" was adopted in the eighteenth century. But it also affected the situation in Britain, Scandinavia, and the Danube Valley.

We are fortunate in having an extremely comprehensive and competent literature on scrapie in England and Germany over the period 1750–1850 and beyond, and there is also substantial literature from Hungary. The reports tend to lack precise details of breeding policies and mating schedules, and are deficient in information about the age composition of flocks and "follow-up" studies. There are frequent reports that the ram is a critical element in the causation of scrapie, and that the disease can be controlled by using rams from flocks free of the complaint. The following interesting picture has emerged.

Around 1750, the disease was serious in parts of central Europe, the Spanish merino being specifically mentioned. The disease was well-known in many of the migratory merino cabanas (flocks) in Spain, and the shepherds recognised it as a slowly fatal and incurable disease, which they detected at an early stage and immediately killed the animals for their flesh "before they had lost bodily condition." Exactly the same procedure is applied today by knowledgable owners of scrapied sheep. The disease was also causing problems at this time in the French province of Berry.

Between 1750 and 1820, scrapie gave rise to increasing concern in northwestern Europe, occurring widely and sporadically, at times attaining calamitous epidemic proportions. In England the disease was most clearly associated with the Dorset Horn, Wiltshire Horn, and Norfolk Horn, all indigenous breeds. In France, Germany, and elsewhere on the continent of Europe, the outbreaks were predominantly found in certain strains of recently imported Spanish merino sheep, especially from some specific cabanas, notably the Escorial, and also from the fine-woolled, small-boned Electoral merino breed of Saxony and Eastern Germany.

SCRAPIE IN THE UNITED KINGDOM

In England, the greatest concern was shown in Wessex and East Anglia, and the first communication on livestock published by the Bath and West Society (formerly the Agricultural Improvement Society of Bath) was on scrapie in Wiltshire, where the disease "within these few years has destroyed some in every flock around the county and made great havock in many." It had been unknown in the area a generation earlier, remained severe from 1770 to 1810, but became uncommon after 1820. It was serious in the Dorset Horn breed by the 1780s, by which time this breed was well established, with its specialised autumn-lambing system. Scrapie was even more damaging in the spring-lambing Wiltshire Horn breed, at that time the main source of commercial flock sheep, and for breeding and the production of meat and wool in the Salisbury

Plain–Cotswold region. In Somerset and Devon, the disease was still, in 1800, "a most prevalent and dreadful disorder," and memories of the disease were still prominent in Berkshire as late as 1840, although by then it was no longer a major problem. In East Anglia, the problem was first recognised in Huntingdonshire between 1730 and 1760, and later became severe in Lincolnshire, Cambridgeshire, Norfolk, and Suffolk. The principal breed affected seems to have been the Norfolk Horn. Deaths from scrapie ran to many thousands, and it was the main cause for the decline and virtual disappearance of the Wiltshire and Norfolk Horn breeds. The disease was also present in the Dorset, Hampshire, and Berkshire speckle-faced breeds, but was virtually unknown in some others, notably the Southdown, Cotswold, Dishley Leicester, the English-bred merino, Ryeland, Shropshire, Mendip, Portland, Lincoln, and Kent (Romney Marsh) breeds.

Scrapie in England was gradually controlled around the turn of the eighteenth century by deliberately using rams of unaffected breeds on ewes of the affected ones. Attempts to use rams from unaffected strains within an affected breed were less successful because of the difficulties in ensuring that the rams were free of the scrapie trait. Scrapie was thus incidentally partly responsible for the introduction of new breeds arising from the crosses, for instance the Suffolk breed, and for subsequent dominance of scrapie-resistant breeds like the Southdown. There was, moreover, intense interest in the selection and development of the British sheep breeds in the first half of the nineteenth century. The meticulous pedigree keeping and selection that this entailed no doubt revealed any undesirable traits like scrapie. The elimination of affected lines and the widespread use of unaffected breeds led to the virtual disappearance of scrapie from much of England by 1850. Some new breeds like the Oxford Down have remained essentially free from the disease ever since. Scrapie was in any case little known in Scotland before 1850, but it was by then present in certain Border Leicester–Cheviot flocks on the borders of Northumberland and Roxburghshire. But Stockman reports that strenuous efforts were usually made to conceal the disease (as often at the present day), and many sheepmen did not recognise the disease when they saw it. By 1910, however, so many flocks had become affected in the Border counties that an investigation was commissioned from M'Gowan of Edinburgh. Scrapie was confirmed as rampant in the Border Leicester, Cheviot, Scottish Half-bred, and Blackface breeds and occurring also in Suffolks, Oxford Downs, and Mules (Border Leicester–Blackface crosses). Greig reported in 1940 that the disease was still widespread in some Border flocks, and scrapie occurs widely in northern England and southern Scotland at the present time.

In England, Wessex has remained largely free from scrapie in the present century, but in East Anglia the disease reared its ugly head again, becoming especially prevalent in Suffolk flocks. Well recognised in 1920, by 1950 it had become a serious cause of concern. But the breeding programme instituted by James Parry has led to a great improvement in the situation of many flocks at the present day.

The period since 1950 has seen closer veterinary inspection of sheep and many spontaneous outbreaks of disease have been reported, cases occurring in most of the recognised breeds. Severe epidemics have occurred in Swaledale and Welsh Mountain sheep. The Swaledale breeders reckoned even in the early 1970s that their losses had already run into several millions of pounds over a five-year period. Scrapie may also occur extensively in some other breeds, but firm information is lacking and there is still today widespread concealment of the disease.

SCRAPIE IN CONTINENTAL EUROPE

In the nineteenth and twentieth century, scrapie persisted in France and Germany (Table 1) for much longer than in southern England, and the disease has occurred continuously in parts of France to the present day. It seems clear that the main culprits were certain breed lines of Spanish merino, which were imported in large numbers because of the increased demand for the production of fine wool. For instance, the "Short Summary of the Trotting Disease of Sheep as an Epizootic Hereditary Disease," published in Breslau in 1828, considered the role of various races of the imported Spanish merinos and the influence of breeding "in-and-in" and stall feeding. In 1868 the disease was reported by May as one of general prevalence in Germany and middle Europe, being rampant in Silesia. It was still widely distributed in merino flocks in Central and Eastern Europe in 1913, but by 1945 it had virtually disappeared. But so had most of the German sheep, and scrapie was a major factor in their decimation.

There is a similar story from France, and the build-up of scrapie between 1780 and 1810 was attributed to the imported Spanish merino and 25 years of consanguineous breeding, with the scrapie trait being associated with other desirable features such as increased muscular development. In both countries, it was observed that certain merino strains, such as the Negretti and Paular, were free from scrapie, and increasingly rams from these strains have been used for breeding. Nevertheless, economically serious outbreaks were still occurring in France in 1870

Table 1
Scrapie in Europe from 1700 to 1985

	1700–1770	1770–1840	1840–1910	1910–1985
United Kingdom	Observed in East Anglia (1730). Becoming severe later, especially in Norfolk Horns, Hampshire, and Wiltshire Horns.	Very severe in East Anglia and Wessex. Use of outcross rams leads to decline later.	Rare in England, but increasing in Scotland, especially in Border Leicester and Cheviot crosses.	Severe outbreaks in southern Scotland. Blackfaces and half-breds now affected. A later decline, but still sporadic outbreaks today. In England, Suffolks in East Anglia severely affected but recent decline. Major outbreak in Swaledales and Welsh Mountains in 1970s, and many breeds still affected.
France	"Vertige" in Berry.	Severe in merinos at Rambouillet. Widespread outbreaks.	Sporadic outbreaks; milk sheep affected.	Widespread outbreaks in many departments. Much concealment but Ile de France and Pre-Alpes breeds affected.

Germany	Observed in imported merinos from Spain (1759). Believed present earlier.	Severe at times in Electoral merinos. Well-documented outbreaks, as at Frankenfelde and Stolpen.	Gradual decline, but German sheep had been decimated by scrapie in many areas.	Becoming rare, but outbreaks in East German merinos after 1950.
Elsewhere	Common in some lines of Spanish merino. Probably present in the Danube Valley.	Severe outbreaks in Hungarian Electoral merinos. Eventually controlled by outcrossing and slaughter.	Becoming rare in the Danube Valley.	Outbreaks in Hungary and Bulgaria in Suffolks imported from the U.K. after 1950. A veritable plague of scrapie in Iceland from 1940 onwards.

Table 2
Natural Scrapie Recorded Outside Europe (1937–1987)

Country	Affected Breeds	Approximate Date
Canada (Ontario)	Suffolk, imported from U.K.	1938
India	Local Himalayan breed, crossed with imported Rambouillet from France.	1940
U.S.A. (Michigan)	Suffolk, imported from U.K. and Canada.	1947
Australia	Suffolk, imported from U.K.	1952
New Zealand	Suffolk, imported from U.K.	1953
South Africa	Hampshire Down, imported from U.K.	1965
Colombia	Hampshire Down and Dorset Down, imported from U.K.	1969
Kenya	Hampshire Down, imported from U.K.	1970
Brazil	Hampshire Down, imported from U.K.	1977
Japan	Local Breed	1982

There has been one subsequent outbreak in New Zealand, again as a result of an importation from the U.K., and many subsequent outbreaks in the U.S.A. and Canada.

and have occurred from time to time ever since. The disease seems to be confined largely to the area south of the Massif Central, with the Berrichon and Lacaune breeds particularly affected.

In Hungary, scrapie went unrecorded until the early nineteenth century, where again the Spanish merino was implicated; but it was conceded that it was "an ancient illness present before the merinos came." All affected animals and their relatives were destroyed in the Hungarian outbreak, and as a result of their more ruthless but effective policies, scrapie disappeared from Hungary long before its decline in Germany. So much so that an outbreak of scrapie in 1959 was initially recorded as being the first occurrence of the disease in that country.

From most other European countries, either there is no information or else sheep are unimportant. Scrapie has in recent years been reported from Bulgaria, ascribed to an importation of British Suffolk sheep; and, of course, scrapie has been since 1940, and continues to be, a major scourge of Icelandic sheep (see also chapter 7).

SCRAPIE OUTSIDE EUROPE

There are no records of scrapie in indigenous breeds outside Europe prior to the European expansion of the sixteenth to nineteenth centuries. The disease did occur in an Asian sheep stock following the importation in the 1920s of Rambouillet rams from France. More recently, scrapie has been recorded for the first time in Japan and, surprisingly, not so far associated this time with any foreign importation.

Elsewhere in the world, large numbers of sheep were imported from Spain after 1808, following the disruption of the merino flocks by the French Army's advance and in the period of expansion after 1815. Australia's national "merino" flocks of 300 million had a complex derivation. Merinos arrived from both George III's English flocks and from the Cape of Good Hope, but there were many crosses with later importations of various breeds from England, France, and Germany. Yet amazingly, no case of scrapie showed up until 1952, when it appeared in an importation of English Suffolk sheep. Again in New Zealand, with their 25 million sheep largely based on English Romneys, Lincoln, and Southdowns, crossed with Australian merinos, no case has occurred except in sheep imported from the U.K. in the 1950s and 1970s. It is not clear whether the antipodean flocks have been fortunate in the importation of scrapie-free lines of sheep (which seems so improbable in view of the scale of importation, although most of the Spanish merinos that went to South Africa were of Negretti or Paular stock, with a low scrapie prevalence) or whether the environment "down under" is inimical to establishment of the disease. It is noteworthy that scrapie also seems to be rare, if it occurs at all, in the Mediterranean countries, and the disease in practice is confined to the cool temperate regions of the Northern Hemisphere. But there is certainly no obvious reason why this should be the case; no such geographical restriction applies to the related human disease.

In North America the disease has often been ascribed to importations of Suffolk sheep from Britain made shortly before or after the 1939–45 war, and a vigorous eradication policy has been in force ever since. All affected flocks and any known contacts are slaughtered. This policy does seem to have been crowned eventually with some success and there are now few outbreaks of the disease reported in the U.S.A. The Canadians are somewhat less optimistic about progress in the eradication programme, and some U.S. outbreaks may have been concealed. Minor outbreaks in Africa and South America are shown in Table 2. All arose as a result of importations from Britain.

HISTORICAL SUMMARY

Thus to summarise the historical position with respect to scrapie in sheep, the main features are:

1. A wave of high prevalence between 1750 and 1820 in England, France and Germany.
2. A rapid decline and virtual disappearance of the disease from southern England after 1820; but a steady increase in disease incidence in Scotland in the nineteenth century, leading eventually to a reappearance of scrapie in England, particularly on the northern hills.
3. A slow decline between 1825 and 1875 in France and Germany, continuing after 1875 in Germany until it had virtually disappeared by 1925.
4. Continuing sporadic outbreaks of scrapie in France up until the present day, and a patchy localised increase in parts of England and Scotland in the twentieth century, culminating in the major epidemic amongst Swaledale sheep during the 1970s and the high incidence in East Anglian sheep of the Suffolk breed at least until very recent days.
5. The failure of scrapie to establish itself in most countries to which European sheep were exported, with the exception of North America. (See Fig. 3.)

It should be added that contemporary opinion in the eighteenth and nineteenth century was that the disease was

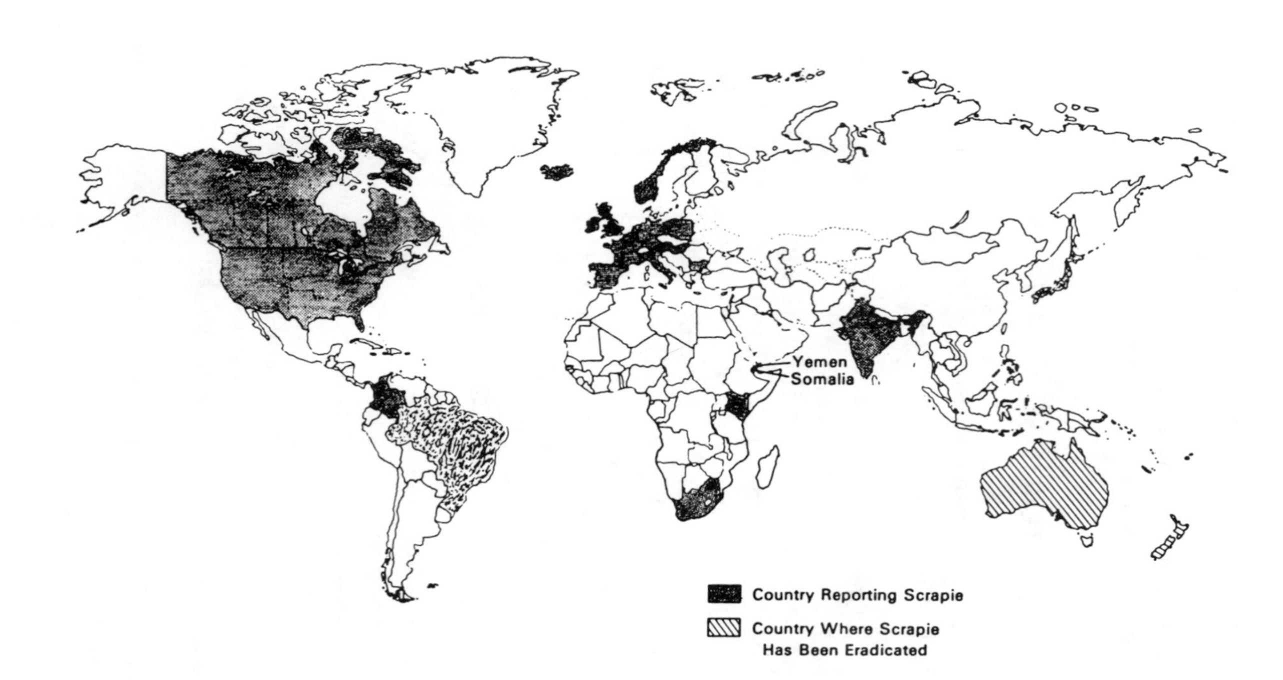

Fig. 3
Scrapie: World-wide incidence.

probably present in Germany and Central Europe before 1750, but that the introduction of the Spanish transhumant (migratory) merinos after that time was associated with a marked increase of the disease, especially in the descendants of the finest-woolled strains of the Electoral and Escorial merino breeds. The practice of close inbreeding with rapid population increase to make the most efficient use of the imported Spanish sheep was considered to be an important factor in the generation of the greatly increased prevalence of scrapie. England recovered more quickly than France or Germany, probably because the importations of Spanish merinos were smaller and selective breeding against the scrapie trait started earlier, the incidence of the disease in indigenous sheep having been in any case probably higher than in France and Germany.

TERMINOLOGY

So far I have, for simplicity, referred to the disease by its currently generally accepted name "scrapie." However, if any reader is desirous of delving into the old literature, then it is necessary to be familiar with the varied and interesting early nomenclature. In the first known account of the disorder in the English language around 1730, it was termed "rickets," but by the end of the century "goggles" or "rubbers" were the preferred terms, along with "shewcroft" and "shakers." "Scratchie," "cuddie trot," and "scrapie" were used in Scotland later; and, as usual, the Scots won their Bannockburn in the end and the latter term became generally adopted, although some shepherds to the present day call it "the plague," just as German shepherds used to call it "die Seuche" (the disease).

In Germany, Leopoldt in 1759 and Erxleben in 1769 called the disease "der Trab" (the trot); later it was referred to as "das Drehenoder Traben" (turning or trotting). Other names included "die Traberseuche," "die Grubberkrankheit" (nibbling disease), "die Reiberkrankheit" (rubbing disease), "die Schefrauderkrankheit" (sheep mange disease), "die Wetzkrankheit" (whetting/honing disease), "die Zitterkrankheit" (trembling disease), and "das Schruckigsein" (shrugging). However, just as "scrapie" won the day in the U.K. and eventually in most of the world, "die Traberkrankheit" (trotting disease) became the dominant term in Germany until the disorder virtually disappeared around 1925.

Scrapie was rare and ill-defined in France in the first half of the eighteenth century and there was in fact no clear nomenclature until Tessier, the superintendent of the Royal Rambouillet Merino stud flocks, introduced in 1810 the terms "la maladie convulsive" and "la maladie folle" (the mad disease). The term "la vertige" (staggers or dizziness) had been used earlier and regarded as synonymous with "die Traberkrankheit." Later, "la tremblante" or "la prurigo lombaire" became the usual terms, and the former is still the preferred name by French research workers.

In Spain, the terms currently used are "la trembladera" and "la enfermedad trotoria," while the most usual term in Hungary is "surlokar" (brushing disease), with "ugeto-myavalya" (trotting disease) reserved for the ataxic form. Finally, in Iceland, the name "rida" continues to be used for the disease, which is undoubtedly identical with scrapie.

Looking to the future, the bureaucrats at the United Nations Organisation, WHO/FAO, have decided that the preferred term is to be "paraplexia enzootica ovium"!

"Scrapie," "la tremblante," "enfermedad trotoria," and "prurigo lombaire" are listed only as secondary terms. However, as the bureaucrats are unlikely to publish anything original on the subject, there is no reason why we should get paraplexed about it.

THE MODERN ERA

The scientific history of scrapie is much shorter, and no laboratory work of any real value was done until fifty years ago when Cuillé and Chelle, in France, showed that the disease was caused by a transmissible, filter-passing agent displaying physical and chemical properties that they considered out of the ordinary. They used the unusual and unattractive route of inoculation of suspensions of brain material from scrapie animals into the eye. Scrapie disease was transmitted and their findings were confirmed in spectacular fashion by the Scotsman, Bill Gordon. He developed a vaccine against another sheep disease called Louping Ill. As with many vaccines, the virus had been inactivated with formalin. Some of the sheep whose brains had been used to prepare the vaccine had been grazing pasture previously occupied by sheep that subsequently developed scrapie. As a result, about a year after the distribution of the second batch of the new Louping Ill vaccine, scrapie suddenly appeared in several locations where it had been rare or unknown for many years. Bill Gordon possessed a remarkable and dynamic personality, and he managed to convince many of the unfortunate farmers that they had helped in the making of a great new discovery. Indeed, this was one of the first indications of the robust nature of the scrapie agent and its indestructibility under many stringent conditions such as prolonged exposure to formalin.

It was very difficult to study scrapie adequately in its natural host, the sheep, first, because the incubation period of the disease is long and variable and, second, the proportion of the animals in any random group being studied that are susceptible to the disease is unknown. In recent years, workers at the Institute for Research on Animal Diseases (IRAD) at Compton, Berkshire, and the Moredun Research Institute near Edinburgh have, by expensive and lengthy selection processes, produced flocks of Cheviot, Herdwick, and Swaledale sheep that can be classed as susceptible or resistant to scrapie under defined conditions. But major scientific advance had, in practice, to await the successful transmission of scrapie disease to other species, where experiments could be carried out more reliably and cheaply. The results emanating from these successful transmission experiments are described in the ensuing chapters of this book.

Chapter 3
Scrapie and Its Cousins in Man and Animals

GOATS AND MICE

Scrapie leapt into the centre of the scientific stage early in the 1960s. The reason was, as so often in science, the convergence and codevelopment of two previously separate pieces of work that greatly enhanced the significance of both. In the first place, research on scrapie had resumed under Bill Gordon at Compton, after a lapse of some years due to the war. With his colleagues, Iain Pattison and Geoff Millson, he showed that scrapie was, often after a time lapse of 2–3 years, transmissible from sheep to goats. Subsequently, Pattison and Millson made a detailed study of experimental scrapie in goats, using some hundreds of animals in upwards of fifty experiments over the period between 1955 and 1965. They were joined by a young American scientist, Bill Hadlow, who made a thorough in-depth study of the pathology of scrapie in the goat.

On the clinical side, they found that individual goats varied greatly in the exterior signs they displayed as a result of the nervous disorder within. Thus, while some animals became hyperexcitable, others were dull throughout the course of the disease. Some animals were easily handled, others were terrified at the approach of humans; some remained active and well-balanced, others showed pronounced abnormalities of posture from an early stage. All in all, the variations were fully reminiscent of the situation long known for the related natural host, the sheep.

But there was a difference. After a few passages (i.e., sequential transmission of the disease from goat to goat), it was observed that two quite distinct clinical syndromes appeared in the goat: the "scratching" and the "sleepy" scrapie goat. The scratching goats first recorded showed areas of broken hair produced by scratching with their horns or hindhooves, and normal "grooming," nibbling, and scratching tended to increase in apparent response to skin irritation. With further disease passage, a far more intense scratching syndrome appeared. The first area of skin to be affected was usually at the base of the neck, occasionally the top of the head. Scratching of the area with the hind hooves usually became so persistent that the skin was broken and bleeding occurred. This period of intense irritation was then succeeded by a less frantic but more general phase, with scratching and rubbing over the rump, along the flanks, and on the sides of the neck. A scab usually formed over the broken areas of skin, which then usually became less sensitive.

The sleepy goats, on the other hand, seldom scratched themselves. They were dull and lustreless, unsteady on their feet, often showing head tremors, reluctance to move, and with obvious derangement of the central nervous system. But as the disease progressed, the end results of unsteady gait and, finally, inability to stand were the same in both forms of the disease. The really interesting point to emerge, however, was that injections prepared from the brain of scratching goats almost always produced scratching scrapie in the injected animals, whereas sleepy scrapie brain when transmitted usually led to the appearance of sleepy scrapie again. This was the first clear evidence that different strains of the scrapie agent existed, and this, of course, had important implications when considering its nature: it was behaving in this respect somewhat like other viruses. It further became clear that the average incubation period was greatly reduced after further passage of the scrapie agent within the goat, a finding that has subsequently been shown to be typical of the situation resulting from the adaptation of the scrapie agent to a new host. However, goats are usually highly outbred, and the incubation period was still long and variable, sometimes years, even after direct inoculation of the goat-adapted agent. Nevertheless, goats were essentially 100% susceptible to the disease, and one could be reasonably certain that a batch of inoculated goats would all succumb eventually.

Even more significantly, Dick Chandler, also working at Compton, showed in 1961 that the disease could be transmitted to mice. Once scrapie was adapted in a suitable strain of mice, the animals were not merely susceptible, but the incubation periods between inoculation into the brain and the appearance of a clinical disease could be as little as four months and were constant for the strain of mice. As with sheep, the mice showed variable symptoms, but Chandler's original description of mouse scrapie in the *Lancet* still stands as typical of much that is commonly seen. The mice stood with their hindquarters lowered close to the ground, and they were reluctant to move (Fig. 4). The hind legs were occasionally dragged, the mice eventually walking with a stiff and rolling gait. The tail was often held in an unnaturally stiff manner, usually to one side; if curled over a finger, it might retain a circular form for several minutes (a question mark tail for a mysterious disease, Chandler used to say). When held up by their tails, most affected mice brought their hind feet together, in contrast with the splaying of hind legs shown by normal

Fig. 4
Scrapie in a white mouse.

mice. Many of the affected mice had ruffled coats and arched backs, and they tended to lose weight because they fed less actively than normal mice. Again, as in the goat, one saw hyperexcitable mice, some of which scratched, and also lethargic mice. Brain derived from a scrapie sheep of the Suffolk breed caused in mice the development of a scrapie line where the mice all became fat. These differing characteristics of scrapie mice were one factor that helped in the isolation of specific strains of the agent that produced constant behaviour in a given strain of inbred mice. Irrespective of the exact disease syndrome, all animals had similar brain lesions in the form of widespread bubbly degeneration, holes in neurones, and enormous swelling of the astrocyte cells.

Scrapie disease became very predictable when working with pure strains of the scrapie agent in highly inbred mice, and it became possible at the time of disease injection to predict the death of an animal to within a few days, even when the signs of disease would not actually be evident for more than a year. With some mouse "models" of the disease, the short incubation period enabled great progress to be made, and accurate measurements could be made of the amount of a scrapie agent present in infected material. These measurements were made by the classical virological method of diluting the infected material continuously until no disease was produced by injection of the diluted material, a technique not often possible when using expensive animals like sheep and goats. Thus it was possible to show that high concentrations of the scrapie agent appeared in most tissues of the mouse during the long course of the infection, and the pattern observed in the relatively early stages is still reflected in the distribution of the agent observed at the time the mouse dies.

It is plain that the scrapie agent does not encounter defensive mechanisms in the mouse that clear it from the tissues or even effectively arrest or suppress its growth. However, disease was found to become evident only when the amount of scrapie in the brain became very great; it seemed that only the brain was damaged by scrapie infection. The success of the mouse work stimulated many further transmission experiments, and Chandler, Gajdusek, and others showed that scrapie could be passed to several other species such as the rat, vole, golden hamster, and squirrel monkey. In fact, scrapie seemed to be potentially a disease of all animal species. What about man?

KURU AND CREUTZFELDT–JAKOB DISEASE

The second area of research slotted in just there. Two Americans, Vincent Zigas and Carleton Gajdusek, had discovered a remarkable disease buried deep in the New

Guinea jungle. An isolated people, the Fore, still in the 1950s to emerge from the Stone Age, were afflicted by kuru. The "laughing death" (a name derived from the fixed facial expression in terminal patients) accounted in the 1950s and early 1960s for most of the deaths among women and children in Fore regions but was rarer in adult males. It was characterized by paralysis and a shiveringlike tremor that progressed to total incapacity and loss of speech, dementia, and death in less than one year from the onset of the clinical disease. The word *kuru* means shivering or trembling in the Fore language. Except in cases of intermarriage, kuru has never appeared in neighbouring racial groups, even though their ways of life and diet were essentially identical.

It transpired that the mechanism by which kuru spread was undoubtedly contamination of the population during cannibalistic feasts involving ritual consumption of their dead relatives as a rite of respect and mourning. They dismembered the bodies bare-handed and did not wash thereafter. Then they wiped their hands on their bodies and in their hair, picked sores, scratched insect bites, wiped their infants' eyes, cleaned noses, and ate with their hands. Liquefying brain tissue containing, we now know, upwards of 10 million infectious doses of kuru was scooped by hand into bamboo cylinders. Women and children crowding about were more exposed to the kuru-infected human tissue than were the men, who left this butchery of dead kinsmen to the women and rarely ate the flesh of dead kuru victims. But it must be presumed that all children who had close kinsmen who died of kuru and whose mothers took part in the cannibalistic rite were contaminated with kuru. Boys without relatives who died of the disease before they left the women's society to live in the men's house would escape contamination and remain free of the disease.

Thus we can explain the sex difference in disease incidence among the Fore people, but not why the Fore alone (Fig. 5) among the tribes of Central New Guinea were afflicted by the disease. The most likely explanation is that by chance a case of Creutzfeldt–Jakob disease, a human nervous disease occurring sporadically all over the world (see below), appeared among the Fore people. When the affected person died, the custom of ritual cannibalism meant that, in effect, the surviving relatives infected themselves with the disease agent. Because the infection was mainly peripheral, i.e., in tissues remote from the brain, a very long incubation period ensued before the disease manifested itself in the infected relatives of the first victim. This permitted intermarriage and reproduction to occur and the disease to spread before the primary victims died. The process would then repeat itself as long as ritual cannibalism persisted.

One further presumes that the process was facilitated in a population that was genetically susceptible to the form of Creutzfeldt–Jakob agent that developed into kuru, and we do not know whether similar extensive, or even complete, genetic susceptibility exists in other human populations. Happily, ritual cannibalism is now a memory of the past among the Fore people, and only 10–15 patients, all over 30 years of age, still die of the disease each year. These patients had earlier appeared to be quite normal and had probably been incubating the disease agent for over 25 years before the clinical signs of kuru appeared.

The link between the two research fields was made appropriately by an American working at Compton. This was the same Bill Hadlow mentioned earlier and he published a letter in the medical journal *Lancet* in 1959 in

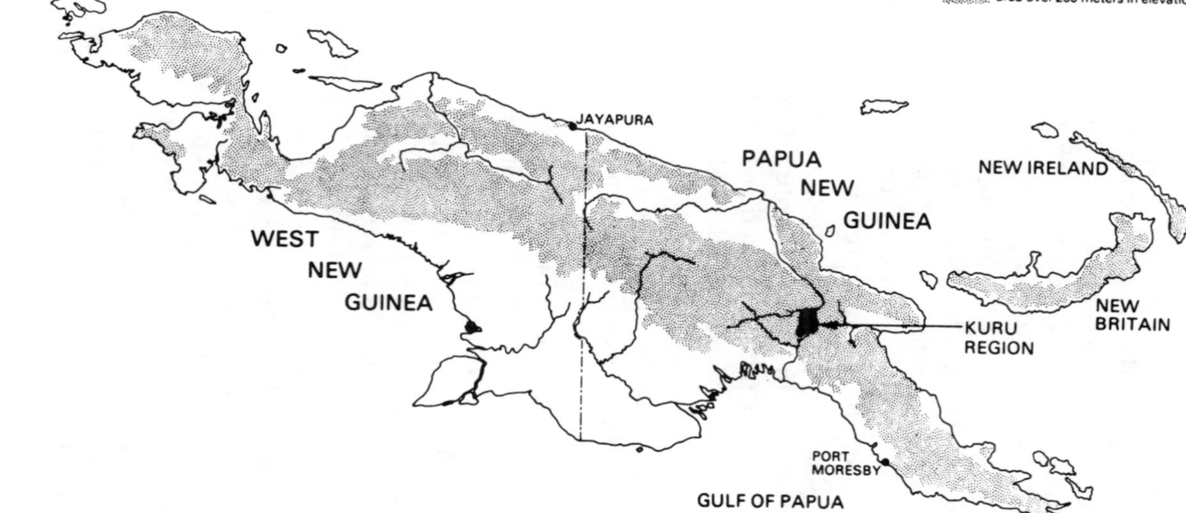

Fig. 5
New Guinea showing the Fore homeland.

which he pointed out the striking similarities between the damage caused to brains of sheep by scrapie and to human brains by kuru. He suggested that transmission experiments of kuru into monkeys be attempted. Carleton Gajdusek had previously thought that kuru was probably a genetically inherited disease, but he reacted with his customary sharpness to the new situation. Knowing that scrapie was clearly a transmissible disease, he took up Hadlow's suggestion and inoculated chimpanzees and other primate species with suspensions of kuru brains into the brain. The chimpanzee was the first to succumb, after about one and a half years, with the same pathological process as that observed in human kuru victims; and subsequently it was possible to establish the disease in several species of New and Old World monkeys. Nonprimate hosts, such as the goat, mink, and ferret, have also proved to be susceptible to kuru.

Closely associated with Carleton Gajdusek in this work was his colleague, Joe Gibbs, who maintained a large research department at the National Institutes of Health in Washington. They realised that kuru was rather a special case and that a more important field of interest in the wider human sphere was Creutzfeldt–Jakob disease. This disease has been reported in most parts of the world, and since the duration of clinical disease is around a year, both the prevalence and incidence of the disease are the same. At any one time, about one person in 1 million is found to be suffering from the disease in advanced human communities; this means that within any generation, about one person in 30,000 dies of Creutzfeldt–Jakob disease of those who live their full potential life span. Thus, although a rare disease, it is not insignificant.

Creutzfeldt–Jakob disease occurs predominantly among people aged between 55 and 75 years, with the peak frequency in the 60–65-year age group; it is thus the most important presenile dementia of man. The average duration of the clinical illness is about 9 months, although many patients do not survive as long and a few live for up to 10 years. However, even before the appearance of symptoms clearly attributable to disease of the nervous system, most patients show some early subjective awareness of physical or mental disorder. This is often accompanied by some degree of muscular or sensory debility, insufficient at this stage to impair normal daily activities. Sometimes a frankly depressive state develops, while other patients suffer loss of appetite or disordered sleeping habits.

The onset of definite nervous disease is manifested both by mental deterioration and by physical signs that progress rapidly. Usually a single symptom or sign gradually initiates the clinical disease, for instance vertigo, episodes of blurred vision or blindness, or the inability to express thought in words. Most common are dementia and behavioural abnormalities, including a broad range of intellectual debilities, from simple memory loss through disorders of judgement and the reasoning processes, to frank states of mental confusion. As the disease progresses, impairment of movement can be seen in most patients, with jerky and trembling movements very common.

As with scrapie and kuru, the brains of Creutzfeldt–Jakob patients show loss of neuronal cells under the microscope and a spongy degeneration with the appearance of holes. A true case of the disease should also be transmissible to animals to produce scrapie-like conditions, and Gibbs and Gajdusek were able to show that Cretuzfeldt–Jakob disease was indeed transmissible to a range of New and Old World apes and monkeys; some strains were further transmissible to the domestic cat, guinea pig, Syrian hamster, and laboratory mouse. The pathology in these animals is indistinguishable at the level of the cell from that in the natural disease or in experimental kuru. Finally, a condition indistinguishable from scrapie was established in experimental goats. This brilliant work, establishing clearly the transmissible nature of a human nervous disease and its relationship to sheep scrapie, deservedly led to the award of a Nobel prize to Carleton Gajdusek in 1975.

The ready transmissibility of scrapie, kuru, and Creutzfeldt–Jakob disease raises many interesting questions and fears were voiced by a few even before the advent of mad cow disease. Is it dangerous to eat mutton, let alone beef? Is it hazardous for nurses to tend patients with Creutzfeldt–Jakob disease or for pathologists to autopsy the brains of patients? The answers, as far as we have them, are, however, reasonably reassuring. It must not be forgotten that most primary transmissions of disease, especially from one species to another, have been effected artificially by the injections of enormous doses of diseased material, often injected straight into the brain. Comparable situations do not often arise in the normal course of events. Creutzfeldt–Jakob disease seems to occur just as often in Australia as in Great Britain, but only the latter country has scrapie as a common disease. A large survey of Creutzfeldt–Jakob disease in France showed that the disease was not associated with shepherds or butchers, although there was a very slight excess of the disease within the medical profession. There does seem to be a greater prevalence of the disease in large population centers, but it is not clear whether this is a genuine result arising perhaps from more opportunities for disease transmission or whether it is a reflection of better facilities for diagnosis of the disease.

There is, however, one disturbing exception to this story. Quite recently, a cluster of about a dozen cases of Creutzfeldt–Jakob disease (or possibly kuru) has been reported from a rural region in Slovakia. All of the patients seemed to have some connection with sheep husbandry, and it does raise the question whether a rare change can occasionally occur in the scrapie agent and allow it to "jump" over into other species such as cows or humans.

Otherwise, the only class of people that do seem to be at risk are hospital patients. There have now been three well-documented incidents of accidental infection of patients with Creutzfeldt–Jakob disease. The accidental infections are usually easy to identify because they result in the occurrence of disease in people much younger than the normal age range for the disease, which in the ordinary course of events rarely affects people under the age of 50. The first recorded accident was in a patient who had received a corneal transplant from a person who had died of Creutzfeldt–Jakob disease. The eye is a rich source of agent in diseases of the scrapie group; and, with hindsight, an accident of this sort, which has since been tragically repeated at least once, was inevitable. The second accident involved the application of stereotactic needles for the measurement of the EEG of patients, the needles having been previously used on a patient suffering from Creutzfeldt–Jakob disease. The intermediate "sterilisation" with formaldehyde was, of course, totally ineffective, and at least two patients in Switzerland contracted Creutzfeldt–Jakob disease in this way. Further cases may still occur. Finally, there have been some very sad incidents arising from the use of growth hormone preparations in young people. Children with pituitary gland deficiencies show retarded growth, but they can develop to the normal adult size range if they are provided with supplementary doses of growth hormone. Until recently, this has been extracted from human pituitary glands, and some of these glands must have been contaminated with the Creutzfeldt–Jakob agent. The unfortunate incidents of Creutzfeldt–Jakob disease that resulted in young people should not be repeated because it is now possible to prepare pure growth hormone synthesised in bacteria by genetical engineering techniques. The worry remains, however, that some other routine hospital manipulations may transmit a dementia, as it is not only the brain of affected patients that contains the agent.

Clustering of cases of Creutzfeldt–Jakob disease in certain areas and at certain times has occasionally been reported. The most spectacular of these is the 30-fold-higher-than-expected incidence of the disease in Israeli Jews of Libyan origin. These people are very fond of sheep's eyes as an article of diet, but scrapie seems to be rare among sheep in Mediterranean regions. The high incidence remains unexplained, but if there is a high genetic susceptibility to the Creutzfeldt–Jakob agent among these people, does this mean that inapparent scrapie-type agents are more common in the environment than the normal disease incidence would indicate?

In a proportion of cases (upwards of 15%), Creutzfeldt–Jakob disease is familial, and in one French family that has been well studied there were probably 14 cases over 3 generations. The disease occurs in a pattern consistent with transmission being controlled by a dominant gene and is therefore a unique example in man of a transmissible disease being also apparently determined by a single gene mechanism. One presumption would be that there is an inherited susceptibility to infection, but more of this later. It is not clear at present how familial Creutzfeldt–Jakob disease relates to the other human nervous disease that has been shown to be transmissible, the Gerstmann–Straussler syndrome (Table 3). This very rare condition is almost always familial, and the dementia is characterised by the deposition of large numbers of unusual forms of amyloid plaques (dense protein deposits) throughout the brain. Somewhat similar amyloid plaques are seen in some cases of Creutzfeldt–Jakob disease, and these patients tend to survive longer than those where they are rare or absent. The survival of Gerstmann–Straussler patients is longer again, often several years, but the disease seems to be very similar to other scrapielike conditions after transmission to experimental animals. It is probably best considered as an unusual variant of Creutzfeldt–Jakob disease.

MINK AND DEER

Two other naturally occurring animal diseases have been assigned to the scrapie group: transmissible mink encephalopathy (TME) and transmissible spongiform encephalopathy of mule deer and elk. The first disease was seen on a ranch in Wisconsin in 1947 and subsequently appeared also in Canada, Finland, and East Germany; the second was first noted in a captive mule deer herd at Fort Collins, Colorado, in 1967. TME was seen again in the U.S.A. in 1961 and 1963 on eight different ranches in Wisconsin and Idaho. The natural disease affects only adult animals retained for breeding because there is a 7–12-month incubation period before it appears and most commercially reared mink are killed at 6 months of age.

Table 3
Slow Virus Infections Occurring Naturally and Caused by Unconventional and Unidentified Agents of the Scrapie Group

Animals

Scrapie in sheep and goats,
transmissible mink encephalopathy (TME),
chronic wasting disease of mule deer and elk,
mad cow disease (BSE),
rare diseases in cats and zoo animals.

Man

Three transmissible dementias:
kuru,
Creutzfeldt—Jakob disease, and
Gerstmann–Straussler syndrome.

All mink ranches are structured in the age distribution of their animals, so that the morbidity in an affected herd will depend upon the time of year at which the disease strikes. Nearly 100% loss of animals has occurred in at least 5 outbreaks. It is always easier to study animals than humans, and by observing the movement of mink between ranches and the introduction of new animals into herds, Hartsough and Burger were able to calculate the minimum 7-month and maximum 12-month incubation period for the disease.

The early clinical signs are typical of a scrapie-like condition: mainly behavioural changes indicative of mental confusion. Aimless circling is often observed and normal patterns of cleanliness disappear, with feed inadvertently walked on. Loss of weight occurs, and the appearance of the animals is changed by matting of the fur and usually arching the tail over the back. When at rest, the mink is unable to maintain the hindquarters in an upright sagittal plane. The animals progress to an increasingly somnolent condition, often standing for hours with heads in the corner of the cage; they become more and more debilitated and die of inanition 6–8 weeks after the onset of the disease.

Burger and Hartsough showed that TME was the first slow virus disease of a carnivore by experimentally transmitting the disease from affected to normal mink. The mink disease is indistinguishable from that induced by inoculating normal mink with sheep or mouse scrapie, and like scrapie, it can be transmitted to many other species such as the squirrel monkey and other monkeys, sheep, goats, and ferrets. It has been suspected in the past that TME arises in mink through ingestion of scrapie-infested sheep meat. Scrapie is known to transmit by the oral route. However, in the early outbreak at least, there was no record of the animals having been fed sheep, only cattle meat. Those who expressed doubts about the universal derivation of TME from sheep scrapie have recently been vindicated. Dick Marsh discovered in 1985 an outbreak of TME in mink that had definitely been fed no sheep meat of any kind.

But the origin of the corresponding disease in the noncarnivorous mule deer and elk is even less clear. Over the 12 years from 1967 to 1979, a chronic wasting disease of mule deer was observed in a wildlife facility run by Colorado State University. At one time almost 80 percent of the animals held were affected and one black-tailed deer also contracted the condition. Clinical signs of disease occurred only after the animals had been maintained in captivity for at least 2 years, whether raised by hand from infancy or captured in the wild as young animals. The disease produced listlessness, progressive weight loss, and depression with insidious onset, and a protracted disease course occurred for up to 8 months, eventually leading to emaciation and death. Behavioural changes included episodes of lack of awareness, and animals would stand for several minutes with lowered head and a fixed stare before reverting to normality. Interactions with unaffected deer decreased, and hyperexcitability with abnormal responses to restraint were also noted.

Subsequently, a similar disease was seen in 6 Rocky Mountain elk that had experienced some contact with affected mule deer. The origins of the disease, which has never been observed in the wild, are obscure in these non-carnivores. They did have minimal contact with cattle, goats, and sheep at the wildlife facility; but these latter animals were not known to be affected by scrapie. However, it is properly classed as a transmissible slow virus disease since it shows the characteristic brain changes under the microscope, and Elizabeth Williams, Stuart Young, and Dick Marsh have succeeded in experimentally passing the disease to healthy mule deer and to ferrets.

We still know little about the survival of scrapie and Creutzfeldt–Jakob agent in nature. In Iceland, sheep were eliminated from certain areas for over a year in the course of the eradication of another sheep disease, visna. When the areas were repopulated with sheep from farms free from both visna and scrapie, scrapie reappeared only on the farms where it had been known before. There is also evidence from Britain for the persistence of scrapie for long periods in pasture and buildings. Again in Iceland, scrapie has been associated with certain human individuals who have been asked, in rare cases, to cease shepherding sheep. But this evidence is becoming anecdotal and, in any case, could relate to pets such as cats or even mice. What is certain is that under appropriate conditions, scrapie-type agents can probably infect most mammalian species and result after long incubation periods in distressing dementias, paralysis, and death. In fact, since this chapter was first written, the disease has been observed in certain other species, such as cats and antelopes; and even more interestingly, there has been a first report of the disease in an avian species, the ostrich.

Chapter 4

The Fourth Plague: Mad Cow Disease

After scrapie, Creutzfeldt–Jakob disease, and mink encephalopathy, a new player has recently emerged upon the stage: mad cow disease. As I write, I can hardly believe that the first case was only diagnosed five years ago. The disease has spread like wildfire since Gerald Wells, of the Central Veterinary Laboratory, Weybridge, identified a sick cow in the West Country as suffering from a subacute spongiform encephalopathy (or spongy brain). Hence the name BSE as an acronym for bovine spongiform encephalopathy. It really wasn't too difficult to make the diagnosis, because when due allowance was made for their size, affected cattle behaved just like scrapie sheep, with most of the variations seen in the smaller animal.

A typical mad cow initially presents with irregular changes of behaviour. As with scrapie sheep, listlessness and restlessness are displayed, and the affected cow will feed more irregularly and separate itself from the rest of the herd to an unusual extent. Cows display very strict hierarchical behaviour. In a herd of, say, 30 animals, all the animals know which is number 14 or, at any rate, numbers 13 and 15 do. If the herd is introduced to a new set of cubicles, number 1 chooses first, number 2 second, and so on; and they keep their places thereafter. Hierarchical sets of this sort are one of the first things to go in mad cow disease, and the resulting disturbances affect the healthy cows in the herd and their milk yields. In fact, if there is a difference from scrapie sheep, it is that cows with BSE tend to be more aggressive. A scrapie ram clearing a five-barred gate is worrying enough, but a ton or so of aggressive cow is quite another thing. Whatever else the problems may be, the handling of mad cows in mid-disease presents severe problems to the herdsman.

The final clinical stage of the disease presents a sad sight, and many readers will have witnessed on television a stricken cow that could only stagger a few steps before falling to the ground. Such animals will, like sheep, survive for quite long periods if tended and fed individually, but, of course, out at pasture they will rapidly succumb to starvation and exposure. In practice, they are slaughtered as soon as the disease can be identified with reasonable certainty, usually a good while before they reach the advanced clinical stage of the disease. But as with sheep, 100% certain diagnosis of disease can only be made by microscopic examination of the brain after death. Did I say 100%? Nothing in life is 100%, and although I have not yet heard of such a case, I have no doubt that, as with sheep, there will be the odd cow where both symptoms and microscopic examination make the diagnosis uncertain.

How did the disease arise in cows? Like so many questions in this area, it cannot be answered with absolute certainty. What is most certain is that it was probably avoidable. When by the end of the 1970s it became clear that many mammalian species were susceptible to the attack of the scrapie agent, it became obvious that the susceptibility of cattle should be examined. Only a few thousand pounds were required at that stage, but "government policy" was against any extension of government research, especially in the agricultural field. It didn't really help if you had a good case, "policy" was what mattered; and in any case, it was difficult to convince the administrators that scrapie was of much importance. After all, the accountants would tell them that it had little chance of making any money within two years! It was certainly a case where a stitch in time might have saved 99, but I'm afraid this has been the story of British science in general for the last 10 years.

But to return to the key question, there are really two theories about the origin of the disease. The most widely canvassed one, which many readers will have heard about on television, is that the disease arose in cattle after they had consumed concentrate food that contained processed sheep tissues. It's difficult to be completely certain of one's facts, but it does seem that around 1980 the animal feedstuff manufacturers started to use more animal waste material than previously; and with the prevalence of sheep scrapie in Great Britain, materials like sheep brain and offal would have been likely to contain large amounts of scrapie agent.

In addition (the accountants again!), an economy seems to have been made in the extent of the sterilisation process. The original process would in any case have been ineffective in destroying all of any scrapie agent present, but the amount of heat used would at least have effected a substantial reduction in the scrapie level. The revised process was even less effective, so that it is a real possibility that cows became infected in the early 1980s from consuming concentrate feeds containing sheep scrapie agent. But it usually requires a very large dose of scrapie agent to persuade it to cross the barrier presented by a new species. Was there enough sheep scrapie agent in the concentrate feeds to enable it to cross the sheep-cow species barrier? We shan't know the answer to this for some years, if ever, because it will require careful typing of sheep and cow agents in mice; and even if someone is prepared to put in the daunting amount of work required, it still may not give a clear result.

This brings me to the alternative and less well-known theory, which in essence is that mad cow disease originates from the cow itself. The point here is that scrapie-type agents may be very widely and commonly distributed, only occasionally causing recognisable disease. There are various suggestive lines of evidence here, for instance the worldwide occurrence of Creutzfeldt–Jakob disease in humans. We know from work with sheep and mice that scrapie can exist and slowly multiply in some animals throughout their lifetime, without them ever showing any sign of disease. Such a situation might also have existed in cows, and indeed the odd clinical case of a scrapie-type disease might well have passed undiagnosed if it was rare. Some veterinarians, with hindsight, believe that they did see the odd case many years ago.

Now it wasn't only sheep material that went into the offending concentrate feeds; cattle products were also included. In this way, a whole range of cattle of varying genetic constitution could have become exposed to a cattle agent, previously perhaps to be found in only a few individuals. Some of the large numbers of cattle now exposed could have had, in analogy with the situation in sheep, a genetic constitution such that the cow scrapie-type agent was able now to multiply rapidly. Since it would already have been cattle-adapted, only a very small dose would have been required to start the process off. Other animals, slightly less susceptible, might have become infected as the agent built up in the environment. I must confess to being one of the minority who thinks this alternative theory actually the more likely to be correct, but it really is little more than a guess.

This leads us on to another question, and the reader will forgive me, I hope, when he finds that once again there are no certain answers. How does mad cow disease spread? Almost nothing is known, but we can move to the sheep scrapie model for some clues. One thing that has been established here is the tendency for the mother to pass the disease to her offspring. Nobody knows how, and the milk seems to contain very small amounts of scrapie agent. I suspect that minor breakages of skin during suckling, especially by older lambs, may be a factor. The other thing we know is that the disease can spread like wildfire within a flock, particularly after the introduction of a new ram, thus altering the genetic constitution of the flock and its susceptibility to a probably preexisting dormant scrapie agent. It has been suggested that consumption of placenta (which contains high levels of scrapie agent in affected animals) might be a major means of spread within a herd. However, an extensive American study in Texas failed to confirm this suggestion, and their conclusion, though tentative, was that scarification, i.e., minor abrasions, provided the main route of entry for the scrapie agent.

The other thing we know from the sheep work is that the scrapie agent is extremely persistent in the environment and impossible to destroy in anything larger than a small laboratory. Conventional methods of sterilisation for bacteria and viruses are almost totally ineffective for scrapie when applied to a farm building. We have the well-documented example in Iceland of the persistence of scrapie agent on farms for periods exceeding one year, and strong evidence for long survival of scrapie agent in pastures in England. There is little reason to doubt that the mad cow agent will have similar persistence.

So, all in all, with cases of mad cow disease now approaching 700 per week, it will be surprising if some cases do not arise by cow-to-cow spread and not merely from consumption of infected food. Will the disease pass from mother to offspring? This is the $64,000 question, and while the balance of the evidence suggests that it will, we must certainly hope that it doesn't. Because, if it does, eradication of the disease will be well nigh impossible with the techniques presently at our disposal; there would be little difference from the situation in sheep, which has proved so intractable.

How dangerous is mad cow disease? Well, one thing is certain: it is very dangerous for other cattle, as the agent responsible is adapted to that species. For that reason alone, every precaution must be taken to identify animals infected with the mad cow agent early on in the disease and dispose of the carcasses as quickly and efficiently as possible. The possibility of other species becoming infected has been highlighted by recent identification of spongy brain diseases in cats, antelopes, Arabian oryx, greater kudu, and eland. However, it seems doubtful whether any of these cases arose directly from the agent of mad cow disease; more likely they arose from consumption of foods related to those that probably gave rise to the disease in the cow or even, particularly in the case of cats, to the greater alertness of veterinary practitioners as a result of recent events. The disease could easily have been present in cats for many years but remained unidentified because numbers affected were low.

So what about the risk to man? Well, the first thing to point out is that hardly anything in this world is hazard free; and since food is almost entirely derived from living organisms, each of which carry potentially harmful germs, no food can in fact be consumed with absolute safety. There are, for instance, significant hazards involved in the consumption of vegetarian diets containing large amounts of fresh fruit and vegetables, which can be smothered with bacteria, viruses, and fungi. Similarly, it would have been ridiculous, even before the advent of mad cow disease, to class British beef, or any other beef, as 100% safe to eat. As I said earlier in this chapter, things are never 100% however much one would like them to be so. But all normal diets can be made *reasonably* safe by taking sensible precautions, such as thorough cooking of meat and careful washing of vegetables.

In the case of mad cow disease, we are dealing with an unknown disease agent. We don't know how it spreads naturally, and, more crucially, we don't know whether we're dealing with a new variant or one of the sheep scrapie strains with which we have lived for centuries. We also don't know how close this agent is to the scrapie-type agents that cause the three related human diseases: Creutzfeldt–Jakob disease, Gerstmann–Strausler syndrome, and kuru. In these circumstances, nobody can possibly say with certainty that there is no potential hazard to man. However, throwing ourselves back on the sheep analogy (because nobody has yet done any real scientific work on mad cow disease, useful government funding having only very recently and belatedly arrived), it seems likely that the risk to man is very small. In the first place, Creutzfeldt–Jakob disease is worldwide in distribution, whereas sheep (and particularly scrapie sheep) are not, nor are cattle (again, particularly cattle with mad cow disease). In the second place, as already mentioned in chapter 3, a ten-year study of Creutzfeldt–Jakob disease in France failed to show any association between cases of the disease and any kind of farming or butchery operation.

Of course, when you are dealing with a disease with a prevalence of only one in a million, one cannot rule out possibilities such as the existence of a few individuals with a genetic makeup rendering them exceptionally susceptible or the spasmodic appearance of a strain of scrapie that would affect such individuals (perhaps the situation in the recent Slovakian outbreak of kuru or Creutzfeldt–Jakob disease). But clearly the great majority of the French population in the ten-year study had not been at significant risk from sheep scrapie. Finally, a third argument against significant risk is the normal requirement for a very large dose of a scrapie-type agent before it can cross the barrier into a new species. If brain and offal are avoided, then other tissues, such as the lean meat cuts, will contain only small amounts of the mad cow agent. It is sometimes stated that they contain none, but this is unlikely to be true in view of what we know from detailed studies of mouse scrapie, where measurement of small amounts of scrapie is relatively easy. In the cow, measurements of small quantities of agent is impossible at present because the species barrier prevents its measurement in mice, and to inoculate cows would be enormously expensive, even if you could find cows that you were absolutely certain were free of the disease. But a small amount of mad cow agent such as is probably to be found in a rump steak might even be better than none. One can often acquire resistance to toxic agents by consuming small doses, and in our present state of knowledge, a small dose of mad cow agent is as likely to be beneficial as harmful. I shall certainly continue to eat and enjoy lean beef until my teeth wear out!

One additional concern that I do have is that the mad cow scrapie agent may be a new strain of agent to which humans have not previously been exposed. If this should be the case, a remote possibility, then some people might be susceptible to it, although even then probably only when exposed to large doses. It would be tragic if a new type of scrapie disease were to affect people at a younger age than those normally affected by Creutzfeldt–Jakob disease. The example of kuru has certainly shown us that under exceptional circumstances such a tragedy is possible. Up to now, we have certainly been more fortunate than some other species in having a scrapie-type disease (Creutzfeldt–Jakob) that is not merely rare but also strikes mainly in late middle age. There is no known theoretical reason why this should be so, but we should certainly try to keep it that way while seeking the elusive cure for people stricken with the disease. We need to keep a close watch on mad cow disease for this reason, but the risk is very small as far as we can tell at the present time. And small amounts of mad cow agent are just as likely as not to actually protect us from Creutzfeldt–Jakob disease.

Another difficult question posed by mad cow disease is how to dispose of the carcasses. The policy of the Ministry of Agriculture has been to dispose of them by burning. Initially, there was extensive burning on open ground, and this has continued from time to time when incinerator facilities have been overstretched. This procedure is, in my view, quite appallingly misconceived. The scrapie agent (and hence, presumably, the mad cow agent) is exceptionally resistant to heat, and a large proportion would simply depart intact with the smoke and gases generated by the fire. I can conceive of few better techniques for distributing it far and wide over the countryside. Of course, the intention is to burn all carcasses in future in modern incinerators with efficient afterburners. Even here I am doubtful whether it will be possible to prevent a small amount of active agent from flashing through the afterburners; the temperature might be several hundred degrees centigrade, but if the agent is only exposed for microseconds it might still survive. But there is no doubt that very little active agent should survive after incineration in efficiently operating modern equipment, and any hazard to man or animals would then be negligibly small. The real risk, however, lies in the fact that in time *all* incinerators break down. Perhaps they wouldn't if a fully qualified engineer was standing by them day and night; but this certainly isn't going to happen day in and day out, and sooner or later the breakdown will occur, with massive release of active agent for up to a mile or so around (farther if there is a strong wind). I well remember our very up-to-date incinerator at Compton breaking down one day, and one could hardly bear the smell of the acrid smoke half a mile downwind of the incinerator.

I consider the ministry's policy here misconceived. What they should be doing is burying the bodies of the cattle in lime on the farms where the disease occurred. If the scrapie analogy holds, the soil surface on these farms will be heavily contaminated anyway, and there would be no danger of contaminating men and vehicles when transporting the carcasses away. There would also be negligible danger of contaminating watercourses underground, because scrapie loses activity if the particles that are active are reduced to a fully soluble form. Although the lime would reduce scrapie activity only very slightly, it would accelerate the disintegration of the carcasses; and scrapie is eventually destroyed by bacterial and fungal attack. It would not survive like anthrax because it cannot form the equivalent of a bacterial spore. Burial in lime on the farm of origin seems to me to carry so many advantages over incineration (it would be cheaper too) that I can only repeat that the Ministry of Agriculture's policy puzzles me.

Where do we go from here in BSE? Well, what we desperately need is more information, and that is going to be slow and hard to come by. I have put forward some informed (but the information limited) guesses in this chapter, and people directly involved in controlling the epidemic will be applying similar guesses. But the really important thing is for the government to adequately fund for many years an adequate research effort by veterinarians in the field and, more importantly (because the key answers aren't going to come from the field), by trained scientists in the laboratory. Regrettably, little advance will be possible without extensive experiments on small animals.

Chapter 5
The Unusual Properties of the Scrapie Agent

"NORMAL" PROPERTIES

A vast range of predators, parasites, and small pathogens (agents of disease) prey upon man and animals. Top of the predatory league is perhaps the tiger, and among the parasites we include things like mites, fleas, and lice, not man's greatest favourites among living things. While it is fairly obvious from what we have already discussed that the scrapie agent must be a candidate for the small pathogen group, we cannot altogether neglect the possible role of larger organisms. Parasites could easily be involved in the transmission of the scrapie agent under natural conditions, and larger animals and plants could provide natural reservoirs for the scrapie agent to survive and multiply away from the species in which it normally causes disease.

Nevertheless, when it is said that the scrapie agent displays unusual properties, we are really comparing it with the groups of microorganisms commonly implicated in disease and especially bacteria, viruses, and fungi. Immediately we hit upon a difficulty: these groups of microorganisms are very diverse in their behaviour and properties. They have all been evolving for much longer than the higher animals and on the microscopic scale are consequently even more variable than the living things we can see with the naked eye. It is extremely difficult to define "normal" behaviour and "normal" properties even for the most uniform of these groups, the viruses, which have the deceptively simple-seeming basic structure of a nucleic acid core surrounded by a protein coat. The reader, therefore, should not take the brief outline below as gospel, and should consult more detailed textbooks if he or she wants a more complete story about the properties and behaviour of microscopic organisms.

Size is an important criterion, and bacteria span a relatively small size range. The largest bacteria are rod-shaped and about 1/200 mm in length; the smallest are tiny spheres with a diameter about 20 times less than this. The smallest known viruses are nearly 1000 times smaller than the larger bacteria, but the largest viruses approach the smallest bacteria in size. Bacteria usually occur as single cells and possess a very simple structure without even a clearly defined nucleus. Yeasts and fungi have a much more complex structure, with various distinct "organelles" such as nuclei and mitochondria present within the cell. They also occur frequently as filaments and mycelia containing many individual cells and are then, of course, visible to the naked eye; but in any case, the individual cells tend to be much larger than bacteria.

Bacteria and fungi can be readily examined in the ordinary light microscope, and the electron microscope is also used to study details of their internal and external structure. Viruses can be visualized only by using the electron microscope, with its enormous powers of magnification. But purified viruses produce extremely beautiful pictures when viewed in this machine, and once identified structurally, they can usually be picked out from pictures taken of the cells in which they are living. Thus we expect to be able to visualise any microorganism, at least in the electron microscope. Killed and dried specimens are normally the subjects of observation in the electron microscope, and some distortion in the picture seen has to be allowed for; but all living things tend to be killed by radiation, especially radiation that is "ionising," i.e., creating electrical charges like an electron beam. Similarly, light such as ultraviolet light, outside the normal range of "natural light" radiation to which we are commonly exposed in large amounts, is very destructive of life. Viruses are more resistant than large organisms such as bacteria and yeasts, simply because they are smaller and present a "target" that is not so easily hit by radiation.

Bacteria and yeasts, again because they exist as whole living cells, are quite sensitive, in general, to chemical attack and heat. They are normally destroyed by boiling, and indeed at much lower temperatures, and solvents and detergents are usually effective in rendering them inactive and provide the basis for many simple antiseptic and disinfectant procedures. Viruses are less complex in structure and grow and multiply only when they enter and parasitise a living cell. They are thus more resistant to physical and chemical attack, and require for inactivation either the disruption of their structure or damage to the individual protein and nucleic acid components. Nevertheless, they can usually be inactivated by heating to moderate temperatures like 60° and few survive boiling for any appreciable length of time. Reagents that combine with their protein or nucleic acid readily destroy the biological activity of typical viruses. The reagent most commonly used is formalin, which often serves to inactivate viruses in vaccine preparations.

On the biological side, the best known property of bacteria and yeasts is their ability to grow, divide, and multiply when provided with the right nutrients. They multiply by splitting in two, and under favourable conditions, this splitting can take place every half hour. Thus, in theory, one bacterium can give rise to 17 million descendants in 12 hours. Bacteria rarely find the necessary ideal conditions outside the laboratory, and if conditions are particularly adverse, some pathogenic bacteria have the power to form spores that have a dehydrated structure, with some properties more like those of viruses than bacteria. Similarly, viruses can also be made to grow rapidly if provided with the appropriate living cells in which to grow; tissue and organ cultures for growing large amounts of virus have become the stock-in-trade of the virologist. Thus, we expect to be able to grow in the laboratory the typical microorganism, although some prove difficult to cultivate in practice.

In the context of scrapie, we are particularly interested in microbial pathogenesis, i.e., their ability to produce disease. The lesions typically produced by microbial attack are inflammation and tumour formation. Inflammation is by far the more common, especially in farm animals that are rarely allowed to live to old age. The word implies heat, and in acute inflammation the affected part is always hot; swelling, pain, reddening, and impairment of function are common accompaniments of the lesion. The swelling that occurs in acute inflammation is due not only to the dilation of the blood vessels but also to the accumulation of fluid in the tissues; the heat is no more than must be expected when blood is flowing freely to the part, because skin temperature is normally well below that of blood. Pain arises because of pressure and is much greater if the lesion is in a confined space, like the hoof of a horse or the human tooth. Inflammation is thus the typical first reaction of healthy tissue to attack by microorganisms; the blood with its battery of defence mechanisms is brought into action.

The major defence mechanism of the blood is the immune system. This is of immense complexity, and the unravelling of its secrets is still occupying some of the best minds among biological scientists at the present day. However, reduced to simple terms, the lines of attack depend upon cells and chemicals produced in the blood and the interaction between the two. The main chemicals are the antibodies that match the surface structure of the invading bacterium or virus and thus can combine with it and lead to its inactivation. In various ways, immune white cells, some known as "killer" cells, ingest and destroy the invading microbes, especially after they have been marked down by antibodies. Again, we expect the typical pathogenic microorganism to evoke an immune response, and the outcome of the disease depends upon the balance between the weapons of the microbe, such as chemical toxins, and the weapons of the immune system, which normally gain the upper hand; otherwise animals would not long survive.

Being constructed mainly of organic compounds, microorganisms are, of course, susceptible to attack by reagents specifically designed by nature to break them down. These reagents are known as enzymes, which can also be mainly concerned with the synthesis of organic molecules. Degradative enzymes, however, can be readily used to destroy bacteria, particularly if they are designed to attack their surface structures. Lysozyme, discovered by Alexander Fleming of penicillin fame, is typical of such enzymes, and enzymes can be obtained in bewildering numbers and variety from living organisms of all types. Viruses, with their simple structure, can be more difficult to destroy, but they are usually sensitive to the attack of nucleases, the enzymes that break down nucleic acids, especially if their protein coat is first destroyed by a protease enzyme or other chemicals. Thus sensitivity to enzyme attack is another property we expect pathogenic microorganisms to exhibit.

Finally, it would be appropriate to mention genetic variation in this brief account of microbial properties. We all know about influenza epidemics. These arise because of the ability of the influenza virus to vary and change, particularly in its surface structure. Natural immunity is lost by the human target and has to be built afresh, and the vaccines developed against the last epidemic are no longer very effective. All pathogenic microorganisms tend to vary in this way, although not all are as effective as the influenza virus in designing a new coat for the next epidemic season. But we do expect them all to show some genetic variation and to exist, at least in a variety of closely related forms.

Well, how does the scrapie agent fare in comparison with "usual" microorganisms?

PHYSICAL AND CHEMICAL PROPERTIES OF THE SCRAPIE AGENT

I have referred earlier to the resistance of the scrapie agent to formalin, a discovery of Bill Gordon's in the 1930s. This is quite striking, and, in fact, slices of scrapie brain stored in concentrated formalin for months still remain highly infective in transmission experiments. It was, however, extremely difficult then to place such observations on a quantitative basis because the sheep available at that time were so unpredictable in their response to scrapie.

I was extremely fortunate to arrive at Compton just as Dick Chandler had established the disease in mice, and we now had an experimental animal that responded uniformly to scrapie, with 100% susceptibility and a long but reliable four-month incubation period between inoculation and the appearance of clinical disease. Dick and I soon showed that quantitative measurements of the amount of scrapie agent in a given amount of tissue could be made both on the basis of the length of the incubation period and by normal viral-type titration, i.e., by diluting progressively until no disease appeared after inoculation. With the help of several skilled collaborators, and notably Geoff Millson and Richard Kimberlin, there followed for me one of the most productive periods of my scientific career; within five years or so, the essential physical and chemical characteristics of the scrapie agent were defined on a quantitative basis.

Initially, perhaps, the most important experiments were related to the size of the scrapie agent, and it soon became apparent that we were dealing with an agent that was clearly of smaller size than typical bacteria or yeasts. Filtration experiments showed, for instance, that as normally prepared from suspensions of infected brain or other tissue, there was a limiting size for material containing active scrapie agent, and it would not pass through filters with pores smaller than about 30 nm in size, i.e., the active scrapie agent did not seem to be a simple soluble protein or nucleic acid molecule but corresponded more closely in size to the small viral organisms like the picornaviruses. These viruses contain a small core of ribonucleic acid (RNA) with a protective coat made up of a few, often just 2 or 3, types of protein molecules. In an attempt to visualise such a virus, Dick Chandler concentrated on the electron microscopy of scrapie, but it was a frustrating time for him. To this day, nothing recognisable as a virion (i.e., virus core and coat) has ever been seen in the electron microscope (EM). Only fragments of smooth membranes are the main components of scrapie preparations as seen in the EM. More recently, amyloid-like fibrils have been observed in the EM but of that more later.

Also at Compton at that time was a very skillful virologist, David Haig, who carried out many valuable experiments on scrapie. His major contribution was perhaps in the field of irradiation of the scrapie agent, which he made in association with Tikvah Alper, who was in charge of a radiation unit at Hammersmith Hospital. Enormous doses of X rays had to be applied to scrapie preparations in order to reduce their activity. Ionising radiation of this type operates on a "target" basis such that a very large molecule presents a correspondingly large target and is more likely to be hit and more easily destroyed than a smaller molecule. On the basis of the results of Haig and Alper, the scrapie agent was deemed to be much smaller than any known virus and corresponded more closely in properties to the plant viroids. The plant viroids are remarkable organisms that have been studied in great detail by Ted Diener in Washington, D.C., U.S.A. He has shown that they consist merely of free, naked RNA molecules unprotected by any protein coat. They are of relatively low molecular weight, often only about 100,000 Daltons (i.e., 100,000 times the weight of a hydrogen atom), a couple of orders of magnitude smaller than the smallest viruses. Thus they compete with the scrapie agent in the Guinness book of records for the title "smallest living thing" and do in fact survive exposure to similarly high levels of radiation. They cause a variety of plant diseases which are presently incurable, the best known being, perhaps, potato spindle virus disease and cadang cadang, which is the scourge of the coconut palm in the Philippines.

Somewhat similar results were obtained when ultraviolet radiation was applied to scrapie preparations. Again, very large doses were required for any inactivation, even when the wavelength of the ultraviolet light was such that it would have been readily absorbed by nucleic acid molecules. The situation is well illustrated in the comparisons shown in Fig. 6. In fact, the scrapie agent was more sensitive to ultraviolet light of wavelengths maximally absorbed by proteins, and these results raised serious doubts about whether the scrapie agent could really be classed as a virus. However, Latarjet, working in Paris, showed that ribosomes, the complexes made up of proteins and relatively small nucleic acid molecules and which are responsible for carrying out vital stages in protein synthesis, display a very similar ultraviolet inactivation spectrum. It could therefore be argued that nucleic acid is not necessarily absent from the scrapie agent but is merely of small size and complexed with protein molecules that, as in the case of ribosomes, are also important for the expression of its biological activity.

The percipient reader will have noted that irradiation of the scrapie agent indicates a size very much smaller than the more direct experiments, such as filtration. This has major implications for the nature of the scrapie agent, which will be considered in more detail in the next chapter, but does suggest a complex structure not all of which is essential for biological activity. Of relevance here is one of my own more striking findings that the scrapie agent is highly hydrophobic or sticky, i.e., it does not like water and is normally to be found associated with membranous debris rich in fat or lipid. We found a particularly close association to exist between the scrapie agent and proteins present in the outer plasma membrane of infected cells, suggesting that these plasma membranes were a major locus for the scrapie agent.

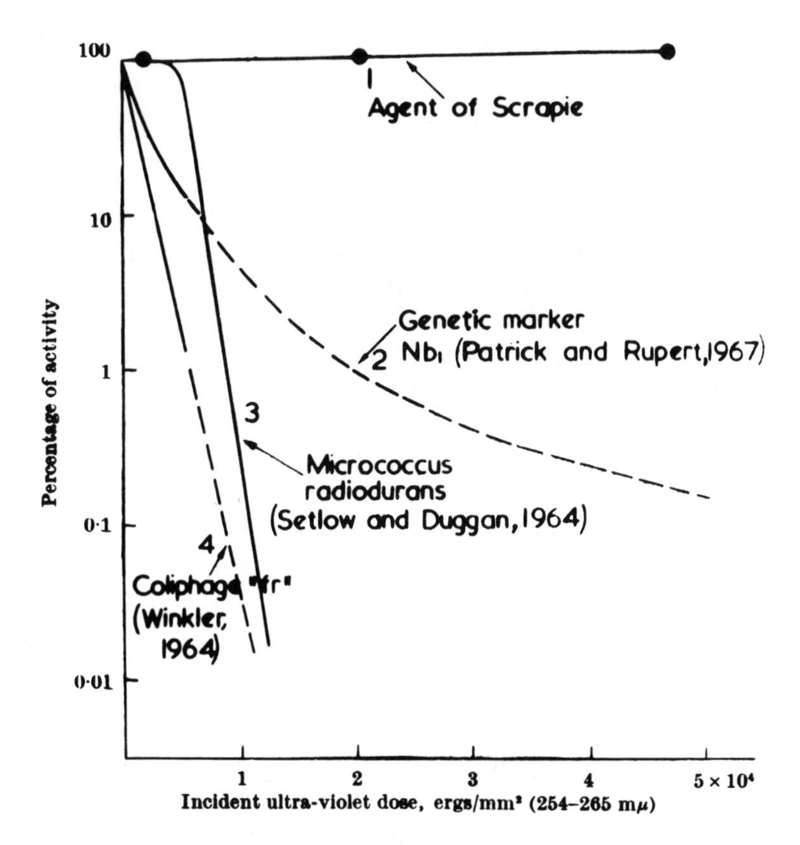

Fig. 6
Insensitivity of scrapie to ultraviolet light.

Another of Bill Gordon's early observation was that infectivity was retained even if the brain was boiled, a quite remarkable finding that set scrapie apart from all viruses known at the time. Using the quantitative mouse assay, we were able to show that although the scrapie agent was indeed resistant to boiling, there was a substantial loss of biological activity and above 121° (the temperature achieved at 1 atmosphere pressure in a conventional autoclave), the activity could readily be destroyed completely by wet heat.

The scrapie agent also proved to be remarkably resistant to the action of a range of solvents, detergents, and enzymes that destroyed most known viruses. Notably, we could not then, nor has anybody since, succeed in reducing the activity of a scrapie preparation by the action of nuclease enzymes, the biological proteins that, as mentioned earlier, break down nucleic acids and destroy the nucleic acid heart of all known viruses. Under certain conditions, however, scrapie activity can be reduced by the action of protease enzymes that destroy proteins. Table 4 provides a summary of the atypical physical and chemical properties of the scrapie agent, most of which were established as a result of work carried out at IRAD, Compton, in the 1960s.

BIOLOGICAL PROPERTIES

With Michael Clarke, one of his colleagues at Compton, David Haig made a further contribution to our knowledge of scrapie. They succeeded in growing infected brain cells in isolated tissue culture that still maintained low levels of scrapie agent over many scores of generations. Many

Table 4
Atypical Physical and Chemical Properties of the Scrapie Agent

Resistant to boiling.
Very resistant to dry heat.
Survives long exposure to formalin.
Largely unaffected by enzymes that attack nucleic acids and other natural molecules.
Resistant to several chemicals, e.g., β-Propiolactone, that normally destroy viruses.
Requires very high doses of radiation for inactivation.
Target size indicated by ionising radiation much smaller than any known virus (about 150,000 molecular weight).
Inactivation by ultraviolet radiation resembles a protein or polysaccharide rather than a nucleic acid.
No virus core and coat recognised in electron microscope.
Amyloidlike fibrils (SAF) seen (sometimes abundantly) in infected brain.

other workers had been unsuccessful in their attempts, partly because the low growth rate of the scrapie agent presents unusual problems. The scrapie agent takes at least 5 days to double in amount in the living animal, and most tissue culture systems are, of course, less efficient than the living animal. On the other hand, the cells that one uses to grow viruses themselves divide regularly at intervals much more frequent than 5 days; so that even if the scrapie agent is growing, it can be doing so more slowly than the cells within which it lives. As the batches of cells are harvested and separated into smaller batches for further growth, they can thus become effectively self-sterilising for the scrapie agent. An unfortunate researcher at one institute actually sustained a nervous breakdown after five years of frustrating and unsuccessful attempts to establish the scrapie agent in tissue culture.

Possibly of even greater importance was the problem of actually getting the agent into the tissue culture cell.

Clarke and others showed that the agent would stay present for months in tissue cultures that were not permitted to grow, but would only multiply when within the cell and then apparently mainly at the time of cell division. Clarke and Haig succeeded where others failed because they used outgrowths from infected brain itself, where scrapie was already growing in brain cells that were themselves capable of only limited growth. This success enabled several interesting investigations to be made, including confirmation of the association of scrapie agent with plasma membrane, but they were still hampered by the fact that the content of scrapie within their tissue cultures was quite low. Nothing would benefit scrapie research more than the arrival of a cell line rich in scrapie, but as with so many areas of biological research, there is currently no way that scrapie can be studied save in whole animal experiments.

Addressing the problems of growing the scrapie agent leads one on to a general consideration of the biological properties and consequences of scrapie and related agents (see Table 5). The major characteristics of the diseases are structural and functional damage to the central nervous system (CNS), with no obvious harmful effects in other organs. If inoculation of scrapie is made into tissues other than the CNS, the agents first replicate silently in the spleen and related tissues before entering the nervous system and brain. They appear to multiply at a very slow rate, with a doubling time of at least 5 days; so it takes many weeks, usually months, for them to reach the high levels at which they cause the functional death of brain cells.

As previously indicated, one of the major characteristics of almost all known viruses is that they present an exterior surface that is foreign to the host they attack. The immune system of the host can then assault this surface in a variety of ways, with cells, protein antibodies, and with

Table 5
Unusual Biological Properties of the Scrapie Group

No remissions or recoveries from established clinical disease; death always ensues.
Long incubation periods can run into decades without any signs of disease.
Very slow growth of agent in experimental models.
No inflammatory response.
No simple immune response; no "foreign" protein detected.
Immune cell function intact.
No involvement of interferon in disease process.
"Degenerative" pathology: protein deposits (amyloid plaques), characteristic fibrils (SAF), and proliferation of certain brain cells (gliosis).
No inclusion bodies seen in infected cells.
Very wide host range suggests that all mammals may be susceptible to a form of scrapie agent.

the complement system of destructive protein enzymes. The scrapie agents evoke no immune response that has been detected, and the course of the disease is not altered significantly in many situations where the immune system of the infected host is suppressed, e.g., whole-body irradiation, treatment with drugs that prevent cell division, sera that destroy white cells, cortisone, removal of spleen and thymus tissues, and the use of abnormal mice with genetically controlled damage to their immune cells. It has been suggested that one of the primary targets for attack by the scrapie agent may be immune cells themselves, the scavenging macrophage cells that wander throughout our bodies devouring intruders being a favoured candidate; but the "Trojan horse" suggestion, though attractive, has never been proved.

When the major nerve cells (neurones) are attacked,

the damage, as briefly mentioned in chapter 1, involves loss of material that shows as microvacuolation, i.e., a mass of tiny holes under the microscope (see Fig. 2a). Nerve cells are like miniature octopuses with large numbers of legs or processes in three dimensions, and the microvacuolation spreads from these processes to the core of the cells. The cells balloon to produce highly vacuolated neurones and eventually spongy degeneration of the gray matter of the brain. The neurones, with their elaborate processes, are made for life and are unable to divide and renew themselves; this is perhaps why they are so vulnerable to the attentions of scrapie. Another type of brain cell, the astrocyte, characteristically undergoes major enlargement in scrapie, and when stained with gold they show up under the microscope as spectacular, large, black spider-like objects. But such an "astrocytosis" is not only associated with scrapie, it is also readily observed in other diseases, for example, rabies; this can be considered a "normal" brain dysfunction.

Most foreign bodies and infectious agents such as bacteria, viruses, and fungi evoke an inflammatory response in their host, with extensive mobilisation of white cells and associated fluids whose task it is to capture and destroy the invading agents. The white cells are transported mainly in the blood stream to their site of action, and as they spread out from the blood vessels, a spectacular "cuffing" of these cells around the blood vessels can often be observed under the microscope. Such perivascular "cuffing" is not seen at all in scrapie. In response to infections of the nervous system, one usually sees a rise in the white cell count and in protein antibodies in the cerebrospinal fluid, but again scrapie produces no such response. Another defense against infection is provided by proteins of the interferon group, which normally mount an early attack on invading viruses. But even in laboratory experiments using large doses, interferon fails totally to check the advance of scrapie. No raised levels of interferons are observed in scrapie animals, although the capability is still there, and further infection of scrapie animals with a conventional virus induces a rapid interferon response.

It would be wrong to give the impression, however, that the characteristics of the scrapie agent distinguish it totally from known conventional viruses. Its strong hydrophobicity or stickiness and strong tendency to aggregate into large clumps associated with membranes could lead to some misleading results when determinations of agent size and assessments of resistance to chemical attack are made. Accurate measurements are also made difficult by the fact that one can only assay scrapie by inoculation into animals and wait for disease to appear. Nevertheless, the filtration data are not inconsistent with the properties of some small, conventional viruses.

Several biological properties of scrapie are typical of the virus world: one would be the ability to grow first in the spleen and elsewhere in the reticuloendothelial system (this is the system of cells found also in various locations such as bone marrow and lymph nodes, and which engulf and break down wornout red blood cells and other debris unwanted by the animal), and later to break through to the brain (rabies does this); another would be the ability of scrapie to transfer to a new host initially with a long incubation period but then adapt to a shorter incubation period characteristic of the new host. The most striking analogies are in the realms of genetics.

Alan Dickinson, working at the Animal Breeding Research Organisation (ABRO) Laboratories in Edinburgh, has shown that many different strains of scrapie exist, with varying abilities to cause disease and differing in host range. These strains have been obtained by passage of tissue from different breeds of sheep through appropriate animal hosts and by selecting strains after enormous dilution (clonal selection) of "wild stock." Dickinson further showed that, as with other well-characterized viruses, an initial infection with a slow-growing strain of scrapie can stop the growth of a normally faster growing strain injected subsequently. These results can be interpreted as indicating that the scrapie agent must contain nucleic acid, changes in which control all the genetic variations that are known about in conventional biology. But there could be other explanations of the genetical results involving, for instance, contributions from the genetic apparatus of the host, and it is time we considered more closely the nature of the scrapie agent, the subject of the next chapter.

Chapter 6
What Is the Scrapie Agent?

EARLY HOPES AND THEORIES

We come to the heart of the matter: the nature of the scrapie agent. It is easy to forget that we have only come to a full understanding of the chemical nature of viruses and the way they are reproduced since the 1939–45 war. To the prewar workers, mostly working before the first virus was fully purified, scrapie was just another virus. But it soon became clear that it was an unusual one, and E. H. Wilson at the Moredun Institute kept the research going in the immediate postwar years. As with Bill Gordon, it was unfortunate that he published little, and many of his early claims, such as the exceptional radiation resistance of the scrapie agent, were discounted because of the lack of any written substantiation of them. However, Bill Gordon and Iain Pattison, who had moved down from Moredun to Compton, were familiar with his work and were early in the field in expressing their doubts about the viral nature of the scrapie agent.

With the availability of the new mouse models for scrapie, a fresh generation of virologists and biochemists took on the problem in the early 1960s, and, of course, none of them paid much attention to the "ravings" of the older generation. I well remember Derek Mould, the chemist who took up the scrapie work at the Moredun

Institute, saying to me in 1963 that we should have the scrapie virus isolated and identified within a couple of years. I had little doubt at the time that he was right. We had used modern methods of cellular fractionation, which involved disintegration of brain cells into their component parts followed by separation from each other in terms of the different rates at which they would sediment when spun at high speed in the ultracentrifuge. Titration of the amount of scrapie agent (see chap. 3) present in the various fractions obtained showed that the scrapie agent could be measured accurately and reproducibly in terms of its biological effect, and as with many known viruses, we found a strong association of scrapie with membranous components derived from the cell cytoplasm (plasma membranes and microsomes) and less associated with nuclei and the soluble cytoplasm. True, we could boil scrapie, but then the nucleic acids of some known viruses could also be boiled without loss of activity, and information on conventional whole viruses with exceptional heat stability was beginning to appear in the literature.

But our troubles now began, and it proved to be extremely difficult to effect any further purification of the scrapie agent. Try as we would, using solvents, detergents, chemicals, enzymes, etc., the scrapie agent obstinately refused to separate itself from cellular debris. To this day, further purification has been very limited. If one went over the top and used extremely potent agents (e.g., phenol or guanidine hydroxide) that were capable of total destruction of the bonds that hold proteins and lipids (fats) together, then the scrapie activity suddenly vanished. Agents like this are often called "chaotropic" reagents, because they chaotically disrupt any ordered structure between large molecules. But they do not actually destroy any of these molecules, so it seemed that they were, instead, breaking an essential link between two components that were both necessary to retain the full biological activity of the scrapie agent.

At about the same time, David Haig and Tikvah Alper carried out their series of experiments on the irradiation of the scrapie agent, with the conclusion that scrapie behaved more like a protein than a nucleic acid when exposed to ionising radiation or ultraviolet light. The older generation were triumphant, and it was felt that here was justification for the earlier (1959) claim by John Stamp, the director of Moredun Institute, that it was "unlikely that the [scrapie] factor would be nucleoprotein in nature." Iain Pattison repeated Stamp's claims in a heretical paper where he emphasised that in his view the scrapie agent could not be classed with conventional viruses. He suggested, on very slender evidence, that it might be a small, basic protein, but that particular claim did not stand up to investigation; basic proteins isolated from scrapie brains under fairly mild conditions showed no biological activity. However, this claim was the forerunner of many unconventional suggestions about the scrapie agent that continue to this day. J. S. Griffith, for instance, made the obvious suggestion that the scrapie agent was more generally a protein, but the conceptual difficulty was that there were no known mechanisms whereby a protein would transmit information back into nucleic acids and reverse the functional direction of the genetic code. Griffith did, however, make some interesting suggestions of mechanisms that might feasibly operate.

At Compton, I was working closely at the time with a brilliant physical chemist, Dick Gibbons. He was an expert on cell surface structure and became fascinated by the apparent strong association between the scrapie agent and cell membranes. Between us we forged the "membrane

Table 6
Membrane Hypothesis of Scrapie
Three Postulates

1. The molecular scrapie agent is closely integrated into a membrane that facilitates its entry into the target cell.
2. Chains of sugar molecules bound to protein (oligosaccharides) may be altered in scrapie.
3. Mechanisms exist whereby these altered sugar chains could be reproduced.

hypothesis" of scrapie, which held sway for a few years (Table 6). We suggested that the scrapie agent might multiply or reproduce as a component of a larger membrane structure. We further pointed out that the short strings of sugar molecules (oligosaccharides) that bind to the protein and lipid (fat) components of cellular membranes (to form what are termed glycoproteins and glycolipids) are, in a sense, genetic materials. Cell enzymes are used to synthesise them, but they can only do so if they are presented with a short chain of sugar molecules to build on. Perhaps one component of the scrapie agent was a slightly different chain structure from normal, which fooled the cell into reproducing the wrong chain. Over the next few years, we were able to produce much evidence that was consistent with this hypothesis in terms of the operational size of the scrapie agent (it was the same as the smallest membranous debris that it was possible to produce by ultrasonic disintegration), its physiochemical behaviour, its intracellular location, and the enhanced activity in scrapie of oligosaccharide synthesis. Over the years the hypothesis has retained supporters, notably Tikvah Alper and, more recently, Dick Marsh; as we shall see later, it holds its place in relation to the most up-to-date research results. There is certainly no doubt that the strongly hydrophobic nature of the scrapie agent makes it associate very readily with membranous material.

SEARCH FOR A SCRAPIE NUCLEIC ACID

We did not, however, at Compton succeed in isolating any single membrane component in scrapie-infected tissues that was clearly different from those present in normal tissues. Scientific attention became focussed in other directions in the 1970s. First, as briefly mentioned earlier, Alan Dickinson succeeded at ABRO in isolating a number of different strains of scrapie agent each with distinct properties in terms of incubation period in a given host and location of damage within the brain. This was the culmination of many years of patient work involving the passage of tiny amounts of scrapie agent isolated from a variety of sheep through various different strains of mice. Some fascinating findings emerged as Dickinson studied the behaviour of these scrapie strains in inbred lines of mice. One was the clockwork precision of scrapie and the remarkably small standard error involved in predicting the incubation period of disease following the administration of a known dose: in some models, with incubation periods of around 500 days, one could accurately predict the time of appearance of clinical disease to within 4 or 5 days. Even more remarkable was the finding previously referred to in chapter 4 that one strain of the agent, called ME7 by Dickinson, had a short incubation period of around 180 days in one mouse strain and a long incubation period of about 350 days in another; whereas, in contrast, another strain of scrapie, called 22A, had a short incubation period of 200 days in the latter mouse strain and a long incubation period of 480 days in the former. Furthermore, in some models it was possible to show that the prior inoculation of a scrapie agent with a long incubation period blocked the replication of an agent with a short incubation period injected subsequently.

All this, claimed Alan Dickinson, added up to close involvement of nucleic acid with the scrapie agent. Certainly, the mice that showed reversal of incubation periods probably differed significantly only in one single operational gene, which Dickinson christened the sinc gene (simply a shortened form of *s*crapie *inc*ubation period gene). He hypothesised that the two alleles of this gene (diploid organisms like mammals can always have their genes in two alternative forms called alleles) produced products that interacted with various forms of the scrapie agent, such that, in his view, the complex variations in incubation period could be explained.

Variations in the scrapie agent itself were not readily explicable except in terms of variation in a nucleic acid component. Well, perhaps. In any case, the biochemists enthusiastically engaged in the search for the putative scrapie nucleic acid. None of us was able to isolate nucleic acids displaying any scrapie biological activity, but several claims for the presence of unique molecules arose as a result of examination of the nucleic acid components of scrapie brain. The researchers knew that any scrapie-specific nucleic acid must either be very small and of unusually simple structure or be closely associated with another large macromolecule. Otherwise, the insensitivity to radiation could not be explained. Claims were made as the result of the work of two young Ph.D. students: "M," working at Riverside, California, and "X," working in my laboratory at Compton.

M was involved with hamster scrapie, and in the usual way she isolated from the brain particulate material rich in scrapie. She carried out some fairly straightforward manipulations involving ammonium sulphate precipitation and enzyme treatment and then fractionated some of her material by gel electrophoresis (i.e., separating materials on the basis of the different speeds they move in solution under the action of an electric current and fixing their movement by adherence to a gel matrix).

We had carried out many similar experiments without success and were surprised by her claim to have isolated a small DNA molecule in one of the fractions. However, in the upshot, all attempts to repeat the work have failed, and it seems possible that M, who was inexperienced, had allowed infective material to drift along the outside of an imperfectly made gel. Similarly, the identification of the active material as a small DNA was made on flimsy evidence based on an incomplete enzyme assay of crude material. Disputes between her supervisors about kudos for the work died down abruptly!

The claims made by X at Compton were more extensive. On the one hand, he alleged that he had isolated a small DNA from mouse scrapie brain and had obtained it in an essentially pure form. On the other, he claimed that there was a large reduction in the amount of one type of RNA (polyadenylated RNA) in scrapie brain from a very early stage of the incubation period. The RNA work was developed in considerable detail, and full papers were published, in collaboration with his immediate supervisor, Richard Kimberlin. This work was so important that, despite X's reluctance, additional workers were drafted to further exploit the findings. Another Ph.D. student, Lynne Bountiff, was unable to repeat the DNA work. At first her failures were put down to lack of experience, but it eventually became clear to me that she was right; and with Geoff Millson's help, she finally produced fairly conclusive evidence for her assertions that the small scrapie DNA of X was nonexistent. Lynne Bountiff subsequently became, I believe, the first person actually to establish a change in scrapie brain nucleic acids, although this was probably not in a component of the scrapie agent itself.

The alarm bells rang, and we now looked closely at the work on polyadenylated RNA. To my horror, two young members of my staff, Alistair Lax and Jane Manning, showed conclusively that none of this work was repeatable either. X maintained for a long time that his colleagues were incompetent, but with the use of blind controls, it became clear that X could not repeat his own work either. X insisted that the experiments had worked previously, but that there was now some unexplained change in the biological systems under study. However, when he resigned suddenly, we decided that further investigations would serve no useful purpose, and X has now made a career in another profession.

This debacle came at an unfortunate time, when the Agricultural Research Council (ARC), now the Agricultural and Food Research Council (AFRC), our paymasters, were attempting to rationalise the research programme in response to financial cuts. Alan Dickinson at ABRO produced, almost out of the hat, an empty building in Edinburgh that he considered suitable for conversion. A committee, under Prof. Peter Wildy of Cambridge University, endorsed his view that this building should become a joint ARC/MRC Neuropathogenesis Unit under the direction of Alan Dickinson, and that all scrapie work should move there. Scrapie work ceased at Compton soon afterwards; the Neuropathogenesis Unit was mainly concerned with its building programme for several years afterwards and was then disrupted by the successive unexpected early retirements of Alan Dickinson and Richard Kimberlin. A committee, reminiscent of so much in British science, had destroyed the lead we held in scrapie, and the British initiative, already being challenged, passed finally across the Atlantic to the U.S.A.

OTHER THEORIES OF SCRAPIE

In the preceding sections, I have described the main-line course of scrapie research. However, along the way many other hypotheses and claims have spawned, metamorphosed, and died. Some of the more interesting ideas, even if unsuccessful in outcome, are worthy of mention at this stage.

The earliest one of these was the claim by James Parry, of Oxford University, that scrapie was basically a myopathy, or at least closely associated with a muscle degeneration. This claim was made in the late 1950s and Parry was perhaps unlucky in that he happened to be working with several sheep flocks that had both scrapie and myopathy at the same time. However, he generously admitted his error when presented with a sheep flock at Compton, where rip-roaring scrapie was well established in the absence of even a trace of myopathic degeneration.

In the early 1970s two groups of workers thought that they had detected scrapie-related factors that affected the behaviour of blood cells. At the Mental Retardation Laboratory on Staten Island, New York, Richard Carp and his colleagues observed that the level of polymorphs (a type of white cell) was lowered in scrapie. They further believed that the factor in the blood responsible for the change could also affect the growth in cells in tissue culture. Was this the scrapie agent? Alas, hopes that their systems could be used to measure and isolate scrapie were soon dashed, as other researchers moved in and failed to repeat the results. Once again, too great a reliance had been placed on the experiments of an inexperienced junior worker, but Richard Carp himself soon recovered from the debacle and has made several thoughtful contributions to scrapie research since then.

Back in England, E. J. Field and colleagues had been staking their claim. They were using the cytopherometer, a delicate instrument that measures the mobility of living cells placed in an electric field. They believed that another type of white cell, the lymphocyte, could be stimulated by scrapie-modified tissues. The lymphocytes were then supposed to emit a factor or factors that altered the mobility of yet another type of cell, the macrophage, in the cytopherometer. It was a complicated system and the changes were not great; so small in fact that other workers failed to detect them reliably. There may have been some substance in this work, but the system was too unreliable to be useful. Interest petered out.

Most interest was aroused, however, by the work of Cho in Canada. Cho had previously been very successful in isolating the virus causing Aleutian mink disease, a great feather in his cap as he had succeeded where others had failed. He was probably overconfident when he decided to take on scrapie. Almost the first the world knew of his incursion was an announcement in the Canadian press that the scrapie virus had been identified by Canadian workers led by Cho, and it was understood that details of the full purification of the agent were likely to follow shortly. At Compton, we soon received details of Cho's work and pictures of particles alleged to be the scrapie virus photographed in the electron microscope. We, like others, found the same particles in normal brain. Combined research with Cho confirmed this, and it turned out that Cho had been so confident that he had bothered to do only limited control experiments with normal brain. He was unlucky in that there genuinely did tend to be rather more of the particles in scrapie brain on some occasions, but they were probably merely particles of the common iron-storage protein, ferritin. Nobody bothered to prove this conclusively once the heat was off, but it certainly was not the scrapie agent. Cho, to his credit, backed down as gracefully as he could.

Before we move to the current front line of scrapie research, there is one more historical item: the Narang particle. There had been similar earlier reports, but Narang gave probably the best description of particles present in scrapie brain characterised by either a circular or rectilinear profile in enlarged nerve cell processes. Dick Chandler collaborated with Narang, who worked at the Newscastle General Hospital, in a detailed study of these particles, which do not seem to occur in normal brain. But unfortunately, they are also often very rare in some scrapie systems. Thus, although they are of considerable interest still, it seems likely that they arise as a secondary effect of scrapie in only some of the disease situations that we know about.

PRIONS AND SAF: THE LATEST CANDIDATES

Fashions change and, not surprisingly, the Americans did not immediately pursue any further attempts to isolate scrapie nucleic acids. Stan Prusiner, working in San Francisco, came to the fore in the 1970s, and has been a dominent figure on the scrapie scene ever since. He reemphasised the hydrophobic nature of the scrapie agent, and concentrated his efforts on Syrian golden hamster–adapted scrapie, where the incubation period was shorter than in the mouse and the amount of scrapie agent in clinical brain at least tenfold more. He made slight improvements to the purification procedures for scrapie in a number of systematic studies, and he extended the range of physiochemical observations.

He was struck by the fact that some of his additional observations did not fit with a conventional virus structure. He further pointed out that the postulated size of the partially purified scrapie preparations left little room for a nucleic acid component. The main component of the scrapie agent seemed to be a hydrophobic protein, which polymerised readily (i.e., adhered to itself) so that the scrapie agent could appear in all sorts of sizes, as indeed it seems to do when one attempts to fractionate active preparations. He coined the name "prion" (*pro*teinaceous *in*fective particle) for the unit of scrapie activity and subsequently isolated protein from scrapie brain that he claimed was the major prion protein. This protein does polymerise to produce beautiful rod-shaped structures, easily visualised in the electron microscope (Fig. 7).

Before continuing the prion protein story, we must cross the American continent to New York. Here, in the early 1980s, Henry Wisniewski and Pat Merz, working at the Institute for Mental Retardation on Staten Island, discovered a new component in scrapie brain: the scrapie-associated fibrils (SAF; see Fig. 8). SAF are distinct particulate structures first observed in scrapie-infected mouse brain preparations by negative stain (i.e., staining the background) electron microscopy. They are composed of two to four helically twisted filaments, each filament 4 to 6 nm in diameter and of variable length; they can be partially purified by cellular fractionation techniques. The SAF

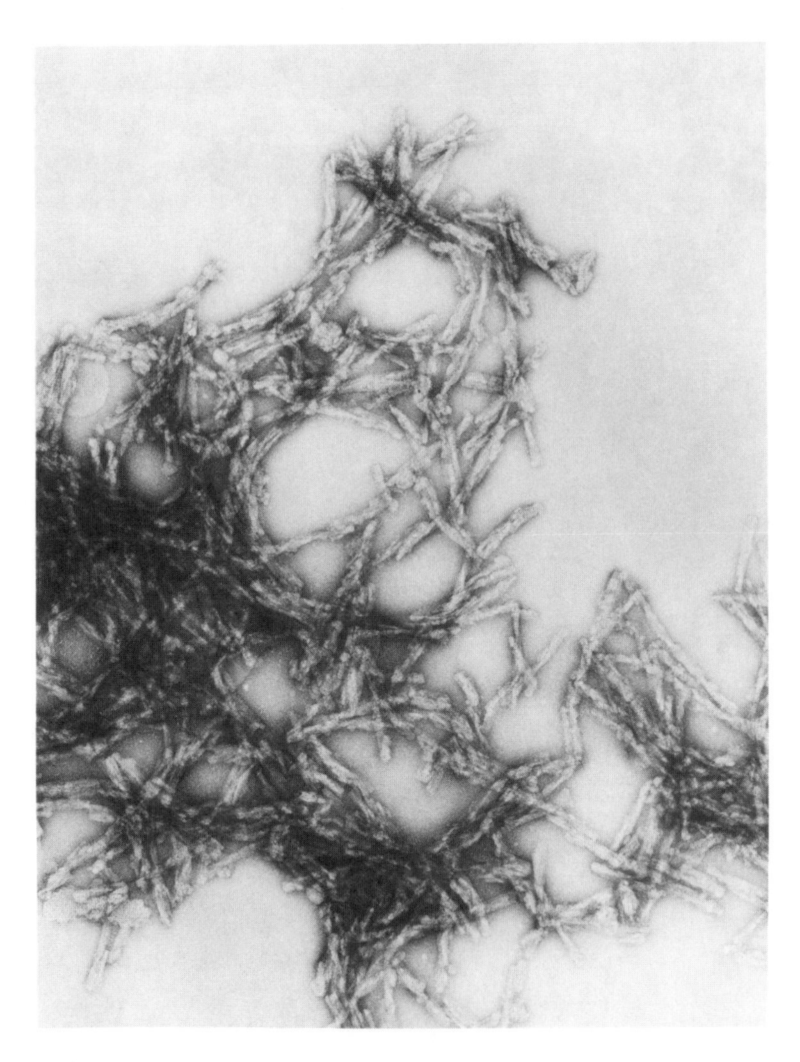

Fig. 7
Prions derived from scrapie hamster brain (magnified 100,000x).

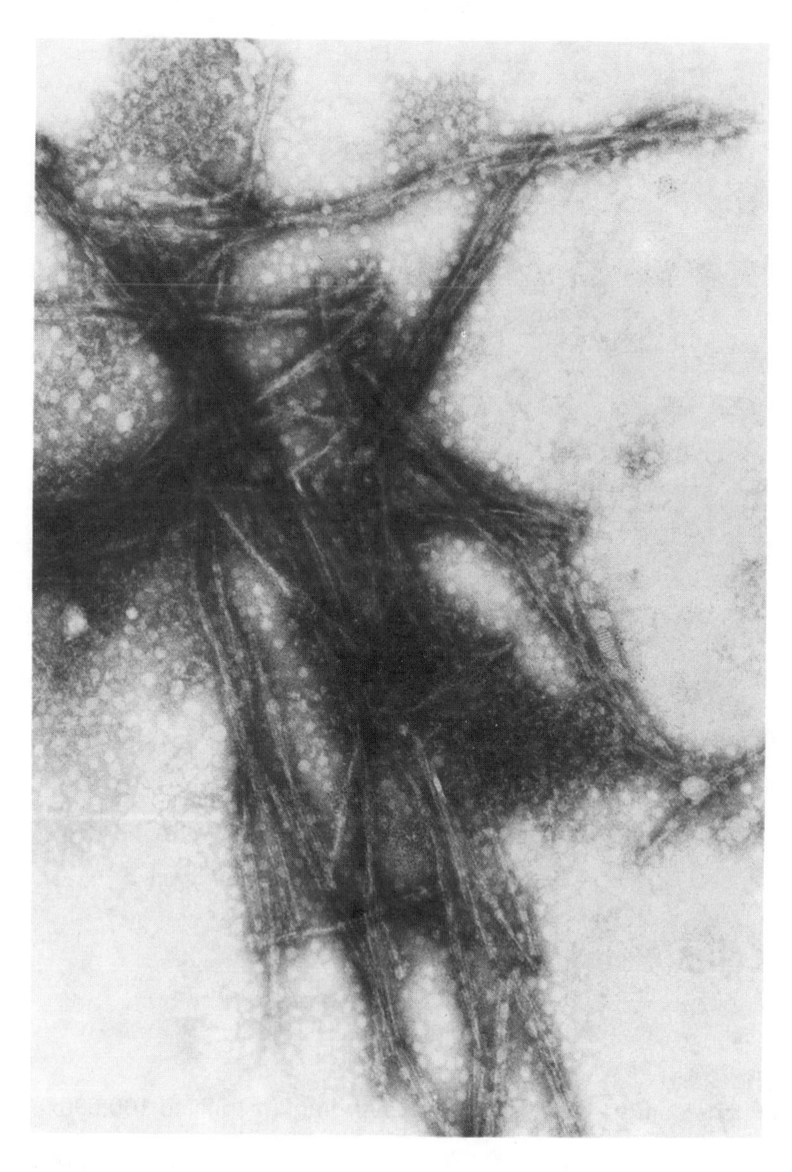

Fig. 8
Scrapie-associated fibrils (SAF, magnified 100,000x).

were considered to be absent from normal brain (certainly rarely found), and although distinguishable, SAF are very similar to structures observed in degenerative brain conditions such as the human senile dementia, Alzheimer's disease. Heino Diringer, working in West Berlin, made similar findings and was in the forefront in developing methods for the isolation of SAF. It was now claimed that the SAF were probably composed of prion protein, which indeed readily polymerised to produce structures very similar to the SAF.

Similar SAF are seen in kuru, Creutzfeldt-Jakob disease and mad cow disease; and Carleton Gajdusek has recently suggested that, in fact, many degenerative diseases of the brain, which variously produce SAF, neurofibrillary tangles, and amyloid plaques, may all have a similar mode of pathogenesis. Neurofibrillary tangles are composed of paired helical filaments, either seen as long, tangled structures or as short pieces 16–18 nm in diameter; amyloid plaques are composed of fibrils of a total diameter 4 to 8 nm, which associate to form the dense bodies or plaques observed in most aging brain but much more frequently in degenerative conditions like Alzheimer's senile dementia. Amyloid plaques are seen in some scrapie animals but neurofibrillary tangles only in Creutzfeldt-Jakob disease.

In the healthy cell, the 10-nm neurofilament may be the key structure. It is normally synthesised in the nerve cell centre and transported down the axonal process, and huge irregular assays of this filament can arise in degenerative conditions and spill out of the cell and evoke an autoimmune response in the animal. In the case of the slow viruses, Gajdusek suggests that they might possess a neurofilamentlike protein component. This would turn on, in neurones, the production of abnormal neurofilaments by distorting copying of the genome, which the neurofilamentlike protein depressor directs. However, it has recently become clear that the prion protein is distinct from neurofilament, or Alzheimer protein, so that the distortion would have to be very considerable; but Gajdusek's general idea may well have substance even if the complete picture requires some modification from the original postulate.

Speculation apart, how close are we now to the scrapie agent? At first, the prion protein was a strong candidate; but when Stan Prusiner showed that it was very closely related to a normal cell protein that is specified by a normal cell gene, it looked as though we were once again on a false trail. The SAF that contain the same protein are still favoured as a candidate scrapie agent by some workers, but Heino Diringer has presented good evidence that the SAF are rather pathological products of infection that can be dissociated from scrapie activity. But scrapie has presented us with a fresh paradox: although the prion proteins from normal and scrapie brains seem to be identical in amino-acid sequence and are specified apparently by the same gene, they nevertheless behave differently when extracted from brain. The normal protein (which is also present in scrapie brain alongside the scrapie version) can easily be broken down to its constituent amino acids by protease enzymes, but the scrapie version cannot. The scrapie prion protein is reduced slightly in size by the enzymes but still retains biological activity when polymerised into rods or associated with membranes. When fully purified, it does lose the ability to cause scrapie disease, but it is then denatured from its natural conformation.

So what is the difference between the normal and scrapie prion protein? Presumably the modification takes place after the protein is made under the control of the common gene. The "orthodox" scrapie workers suggest that the modification is likely to be binding to a small, scrapie-specific nucleic acid to form a complex that Dickinson calls a "virino." The binding would be presumed to alter the prion protein conformation so that it could resist the attack of protease enzymes. But the snag is that, despite renewed extensive searching, no one can find the Holy Grail of the scrapie nucleic acid. Analytical techniques are now so good that, for it to have been missed, it would have to be both very small and 100% infective (few viruses achieve 1% infectivity).

What is the alternative? Our old membrane hypothesis rises from its ashes with renewed vigor. In the first place, the prion protein is a membrane protein with no less than 2 hydrophobic regions that take it through the cytoplasmic membrane of the cells. Secondly, it is highly glycosylated (i.e., it carries extensive oligosaccharide sugar chains). It thus fulfills all the requirements of the membrane hypothesis as a potential carrier of information. Research workers in scrapie are currently busily engaged in detailed comparisons of the sugar components of normal and scrapie prion protein. An additional advantage of the membrane hypothesis is that it would take care of one objection to the potential key role of the prion protein in scrapie. Laura Manuelidis and others claim that there is not always a good correlation between scrapie biological activity and prion protein content. But under the membrane hypothesis, other proteins could, in certain cells and tissues, take over from the prion the role of oligosaccharide carrier, and so remove this difficulty. In summary, then, the membrane hypothesis is alive and well again, almost a theory now, and the soul of Dick Gibbons still goes marching on into the 1990's.

One further recent discovery of great interest is that one of the protein components of the AIDS virus is similar to the prion protein. AIDS virus is now known to damage the central nervous system, and in this respect closely resembles the related visna virus in sheep. Perhaps all three of these agents—scrapie ,visna, and AIDS—affect the central nervous system in a similar way through the action of a prion-type protein. Probably this finding does not further the real advance of scrapie very much, but it does raise the question of whether, perhaps, AIDS virus arose as a result of a human virus picking up from sheep a gene involved in scrapie or visna disease. However, the answer is probably no, because even more recent work has cast some doubt on the closeness of the relationship between the AIDS protein and the prion protein.

Returning to scrapie, the search goes on, and it seems incredible that in these days of advanced molecular biology, we still cannot fully identify something that can be present to the extent of almost 1 trillion infectious units in a single hamster brain. The problem seems to be that scrapie activity is generated by a multicomponent system. We know of extreme cases of such systems among plant viruses, which can have their genome, all of it essential for reproduction, packaged up in several separate pieces. The prion protein is now a strong candidate for one component; but as we have seen, the second component could be nucleic acid, sugar chains, or something else altogether, such as conformational restriction imposed by specific location within the membrane. It might even be a small regulatory molecule operating on the cellular genome. The reader is entitled to a guess, which might be as good as mine, but I suspect that the time for guessing will soon be over.

Chapter 7
Personalities in Slow Virus Research

INTRODUCTION

Two events heralded the advent of slow virus research on the biological scene: one was the first transmission of scrapie by the Frenchmen Cuillé and Chelle; the other was one we have not mentioned before, the importation in 1933 of a batch of German Karakul sheep into Iceland. The flock from which the Karakul sheep came still exists in Germany and is healthy. What happened in Iceland is a warning against the careless introduction of new genetic stock into an unrelated group of animals. Within a few years Icelandic sheep, which had been isolated for centuries, became prey to a whole range of diseases and, in particular, scrapie (or rida in the Icelandic tongue), visna-maedi (the sheep equivalent of human AIDS), and jaag-siete, a transmissible cancer of the lung. Apparently, the native Icelandic sheep had little resistance to these diseases.

Our first personality cut through the Gordian knot of confusion and identified the different diseases and their origin. I never had the privilege of meeting Bjorn Sigurdsson, but I did meet his colleague Pal Palsson. Sigurdsson seems to have been a remarkable man, and in an outstanding lecture given in London in 1956 and in his other writings, he was responsible for initiating the new field of slow virus research. He pointed out a number of common properties among slow viruses, but today we would distinguish between the "slow" viruses like scrapie, which genuinely multiply slowly, and the agents like visna-maedi, which can multiply quite quickly but only slowly damage their target tissue.

Pal Palsson, Halldor Thorma, and others continued Sigurdsson's work in Iceland. Visna-maedi has now been eradicated by the application of a ruthless slaughter policy (not possible for human AIDS!), but scrapie or rida (the Icelandic name means shaking or quivering) persists. It is a tough nut to crack.

COMPTON AND MOREDUN

The stage shifted to Britain, and a number of colourful personalities moved in to occupy it. Bill Gordon, the director at Compton, was notable among these. As we mentioned earlier, he had inadvertently been responsible for spreading scrapie around the country with a batch of Louping Ill vaccine. As soon as the opportunity arose in the mid 1950s, he resumed scrapie research at Compton. Not for the first time he came into conflict with administrators at the ARC head office, who (wrongly most would now agree) took the view that scrapie was unimportant and that research on the disease should be discouraged.

Bill Gordon was not easily discouraged: he obtained funds from the U.S.A., where they were concerned to eradicate scrapie from their country and prevent further importation of the disease. He further requested that the

ARC head office authorise the purchase of 3 dozen individuals from each of 24 different breeds of sheep to investigate their differential susceptibility to scrapie. The ARC head office refused. Bill Gordon purchased the sheep from local farm funds. The ARC head office ordered the sale of the sheep. Bill Gordon ignored the order and inoculated all the sheep with scrapie. There was not really much that the ARC head office staff could do unless they dismissed Gordon, and of course they drew back from anything so controversial. However, they did starve the Compton Field Station (as it was then called) of funds, and only as Bill Gordon's retirement approached did they authorise the overdue conversion of the field station to the Institute for Research on Animal Diseases and brought in David Haig to start a Department of Virology and me to build a Department of Biochemistry.

I found Bill Gordon a fascinating person. He had two scientific interests, scrapie and grass sickness in horses, and could discuss either endlessly. He had a habit of playing golf all afternoon and on his return calling people up to his office just as they were leaving and holding them in conversation for a couple of hours. I managed to beat him at his own game when I discovered that he dined at 7:30. He never troubled me late in the day again after I kept him talking one night until 9:15 P.M.

The director of the Moredun Research Institute, John Stamp, also initiated a scrapie research programme in the 1950s, and the two directors, both strong personalities, were soon at loggerheads. The situation had reached breaking point about the time I arrived at Compton in 1961; and after taking advice from the American director of the Philadelphia Wistar Institute, Hilary Koprowski, the ARC decided to set up a "Scrapie Working Party," which excluded the two directors and met alternately at the

Moredun and at Compton at six-monthly intervals. It was chaired by a Dr. Scarisbrick from the head office, a gentleman of mild disposition, who did the job well up to a point but was obviously in continuous terror that fighting would break out between the Moredun and Compton contingents, urged on by the directors waiting impatiently outside the committee room door. On the Compton side were David Haig, the pathologist Iain Pattison, and myself; while Moredun was represented by the pathologist Ivan Zlotnik, virologist John Brotherston, and a physical chemist, Derek Mould.

Derek Mould and I, of course, considered that the future belonged to biochemistry and to us, and we always got together before the meeting to agree on a common line. I was a little discomfited to find that any clever idea I thought I had also seemed to have occurred to Derek Mould, and he was a little unfortunate in that because we had the mouse model first, we tended to get our results out before him But in the main it was a happy relationship.

Not so Pattison and Zlotnik. I liked Ivan Zlotnik: he was an engaging Pole who always prefaced any controversial communication with the statement, "And now I tell you ze truth," which always seemed to infuriate Iain Pattison. Iain Pattison presented a very staid appearance to the world, but underneath was a passionate personality who found it very difficult to comprehend anybody else's point of view. Among his other accomplishments, he wrote love stories for women's magazines, and his vibrant clashes with Ivan Zlotnik dominated the early meetings of the working party. Each of them considered that the other was stealing his results, although Mould and I rather disdainfully considered the issues of scrapie pathology of minor importance (our time of trial was to come), while Brotherston and Haig were mainly concerned with a mutually sympathetic explanation of their difficulties in attempts to grow the scrapie agent in liquid culture, where success would have reduced the expense, complications, and numbers of whole animal experiments.

The working party came to a somewhat precipitate end in 1966 when the rest of the group became critical of some of Iain's more extravagant claims to have purified and made progress towards the identification of the scrapie agent. These claims could not be substantiated, and Iain Pattison eventually retired (at 62, which was then considered early) in 1973 but still convinced that he was the only one in step on scrapie.

DICKINSON AND PARRY

One of Pattison's strongest critics in the later years had been Alan Dickinson of the Edinburgh Animal Breeding Research Organisation. Alan Dickinson has strong views about scrapie and how the problem should be approached. Music, however, is probably his first love, and he likes to write his scientific papers in the small hours of the night to the accompanying strains of Beethoven and Bach. Perhaps this accounts for the complexity of some of his prose. He has been described as the conscience of scrapie research, and certainly he was conscientious in his sustained criticisms of the alleged deficiencies in the work of James Parry at Oxford University. James Parry had been investigating a major outbreak of scrapie among Suffolk sheep in the south of England and had come to the conclusion that scrapie was a genetic disease that could be controlled by selective breeding with what he termed his "Chinese white" rams. He eventually accepted that scrapie could be transmitted experimentally, but he always maintained that this rarely, if ever, happened under "natural" conditions and that the experimental disease, generated by the product of a defective gene, was distinct from the "natural" disease.

James Parry was another scrapie personality that I had a lot of time for, and he was a man of wide culture and extensive interests. A fine monument stands to him today in Wolfson College, Oxford, a successful blend of traditional design and modern materials. He chaired the architectural liaison committee and over several years visited every British university and many similar institutions abroad. He adopted the wise technique of always seeking out junior staff and asking them what was wrong with their building, and Wolfson College benefited as a consequence in many detailed aspects of design.

In scrapie, James Parry could eventually point to the success of his methods in practice in reducing the scrapie problem within the Suffolk sheep flocks, and the Suffolk sheep breeders certainly valued his work and showered him with prizes and commendations. Alan Dickinson, however, accused Parry of using circular arguments to generate unsound science, and a series of classic confrontations ensued at scientific meetings. These culminated in the famous "Battle of Washington" in 1964, also attended by Gordon and Stamp.

The American sponsors were astonished to witness the violent arguments between the British scrapie workers, which included dramatic walk outs and scathing criticism of each other's work. At the time, Dickinson, a skilled geneticist, probably did get the best of the argument, particularly because, as time went on, it became clear that the "experimental" disease in Cheviot, Herdwick, and Swaledale sheep was controlled on a basis of dominance, the reverse of Parry's claim. At first, it did not seem very sensible to give such credence to Parry's claim that the "natural" and "experimental" disease were distinct, especially as it became obvious that there were some exceptions in the classification of sheep by Parry. However, recently an extensive study of scrapie in Île de France sheep has been carried out in French flocks, and this seems totally to support Parry in that scrapie appears to be controlled by a recessive gene.

Looking at the situation in 1992 with the knowledge that we now have of modern molecular biology on the one hand, and of the scrapie prion protein on the other, it does look as though Parry may have been right after all. If the prion protein has a nucleic acid bound to it, this would be a product of the scrapie gene and might be capable of independent replication, provided only that there was an additional protein present in the brain (reverse transcriptase) that was controlled in a dominant manner. It is even easier to explain the "natural," "experimental" difference if the membrane hypothesis proves to be correct, as the scrapie gene product would then automatically be infective because the sugars on the prion protein would induce the increased synthesis of enzyme proteins and cause reproduction of the modified prion protein in much the same way as in the natural disease. In either situation, there would always be the possibility of an admixture of some "experimental" disease within the "natural" disease, thus explaining the exceptions that Parry himself was reluctant to recognise.

GAJDUSEK AND GIBBS

On the other side of the Atlantic, Carleton Gadjusek and Joe Gibbs were at this time engaged in their classic series of experiments demonstrating the transmissibility in the laboratory of the human dementias, kuru and Creutzfeldt–Jakob disease. Gajdusek is a remarkable man with a great breadth of interests; not since Erasmus Darwin can a scientist have managed to encompass such a broad range of subjects. In addition to his work on slow viruses, he can claim to have studied many other tropical diseases, especially in the South Pacific/Australasia area, to have completed the map of the inhabited world, to have studied the culture and languages of several primitive races (in New Guinea and South America particularly), to have made many contributions to tropical zoology and botany, and to even have found time to detect evidence for continental drift in his biological studies. One of the last of the great explorers, he descended by helicopter on tribes in South America that had never previously seen anything of modern civilisation, and he mapped out the location and relationship of the last villages to be discovered in the Amazonian jungle. His exploits in the highlands of New Guinea were even more spectacular. Many of the tribes there were still in the Stone Age in 1950; and as every village tended to be at war with the neighbouring villages, there were no guides who knew more than a very limited amount of territory. Carleton Gajdusek mapped out much of this new territory and recorded customs and the many different languages. A gifted linguist, Gajdusek claims to speak about 37 languages, including all common European tongues, although as Pal Palsson once remarked, "He does not speak Icelandic."

Visiting Carleton Gajdusek's house in Washington is like attending a meeting of the United Nations held in the Metropolitan Museum. In addition to the continual traffic of visiting scientists of all nationalities, he has adopted over the last twenty or thirty years two or three dozen orphan boys from various tribes in the South Seas. You sit down to dine with a boy from the New Guinea Kuku-Kuku just out of the Stone Age on one side of you, a Solomon Islander on the other, and the mathematician who invented Algol opposite. The house is full of artifacts from the South Pacific and elsewhere, on the way to the many museum collections that have been enriched by Carleton's generosity.

To live such a life is very demanding, and I well remember becoming slightly unpopular by declining to drive Carleton down from Newbury to London at six in the morning after we had been up until three o'clock discussing scrapie. He wanted to be at the British Museum in time for its opening at eight o'clock, because the rest of his day included a visit to Keats's house in Hampstead (he was studying the poet Keats at the time), then calling at the Imperial Cancer Research Fund Laboratories at Lincoln's Inn Fields, then to Cambridge, finally a visit to Stonehenge, and a flight back to the U.S.A. the following morning. I gather the view of Stonehenge at midnight was somewhat obscured by mist; I should have made that early morning drive and gone that way round.

Some critics, like Alan Dickinson, have felt that some of Gajdusek's work lacks depth, but few would doubt that he richly deserved his Nobel prize. He was groomed for scientific stardom from childhood and started involvement in research work at the Rockefeller Institute in New York at the precocious age of eleven. It is said that names of the Nobel prize winners were written on the stairs at his aunt's house, but the top stair was always left vacant.

The only regret I have is that Joe Gibbs did not share the prize with him. While Carleton was stumping through the New Guinea jungle, Joe was usually holding the fort in Washington and seeing the complicated series of transmission experiments through. The people who bear the brunt of the work in a successful programme often get forgotten by the Nobel committee. I witnessed a wonderful example of Joe Gibb's persistence when visiting Washington in 1978. While I was there, Gajdusek rang from Indonesia to say that "he had not got round" to obtaining a Russian visa when last home and had to attend a meeting in the Soviet Union the next week. Could Joe fix it for him so that he could gain entry through Vladivostok after visiting Japan? Surely, I thought, Carleton Gajdusek is in difficulties this time. Joe Gibbs rang the Soviet Embassy and, of course, received the stock reply, "It is impossible." Nothing daunted, Joe harangued the unfortunate consular official and threatened to expose to the Moscow Academy of Sciences his bureaucratic hindrance of the work of a great scientist and friend of the Soviet Union. To my surprise, it was the Russian who eventually cracked, after a prolonged argument, and Carleton Gajdusek had his visa duly delivered to his Tokyo hotel a couple of days later.

This was not the first time by a long chalk that Carleton Gajdusek's reputation had worked wonders in the corridors of power. Some years previously there was a good instance of this when Gajdusek had accumulated a vast quantity of bulky historical, archaeological, and artistic material and had arranged to transport it to the coast of New Guinea. But how to get it off? He managed to persuade the French Navy to divert a destroyer to pick it all up for him and transport it to one of the American-controlled islands. I suppose the destroyer was not really doing anything in particular at the time, but few of us would have the nerve to try and divert it—let alone succeed in doing so.

NEWCASTLE–UPON–TYNE

Another British medical scientist to catch the eye at this time was Prof. E. J. Field, director of the Medical Research Council (MRC) Unit of Research into Demyelinatory Diseases at Newcastle–upon–Tyne. He was a dominant individual of impressive physique and strong views, which he did not hesitate to put forward forcefully at any time. Visitors to his unit were often disconcerted by the vigorous arguments that developed at the drop of a hat between E.J. and his staff. E.J. was also singularly fearless and this showed very clearly in his driving. He himself claimed never to have given way until he saw the whites of the other driver's eyes. I myself witnessed one driver go more extensively white in a split second as E.J. cut across him. Competition for the back seats in his car was severe, and I had earlier narrowly escaped being trampled under by Iain Pattison's panic-stricken rush to the rear.

E.J. was determined to make his mark on science before retiring and in a short period of time he related significant advances in his research work to the problems of cancer, ageing, multiple sclerosis and scrapie. Criticisms from Alan Dickinson and others in E.J.'s own unit caused the M.R.C. to stir uneasily. The interesting upshot was that the administrators instituted an investigation of man-management at the unit!

The investigation became hilarious. Clive Jenkins's union representative was bodily thrown out and promptly issued a note calling all the technicians out on a one-day strike. E.J. intercepted the note and tore it into little pieces. The union representative then managed to smuggle a message through to the investigating committee and called everyone out for two weeks. E.J. meanwhile enlisted the support of the Sunday *Times*, and he was featured as the scientist who worked day and night (as he did) and was being pilloried merely for expecting his staff to do a good day's work too. As usual, the issue was fudged, and E.J. was allowed a separate unit of his own (without any appreciable resources). The MRC Unit of Demyelinatory Diseases limped on for a year or two but eventually closed.

BACK AT COMPTON

With James Parry, Iain Pattison, and E. J. Field out of the field, Alan Dickinson became very critical of the membrane hypothesis of scrapie put forward by Dick Gibbons and myself, although he was unable to disprove the main idea that the scrapie agent appeared to be closely integrated with membrane structures. Moreover, he had in Dick Gibbons (G as everyone knew him) up against him a person of powerful intellect, who was more than a match for him in scientific argument. G was also one of the gentlest and most lovable people I have ever met; I do not think I ever heard him say an unkind word about anyone and this was reciprocated.

G compensated for his usual long-suffering kindness by being extremely devious when playing the intellectual games of which he was very fond. We had one famous game of "Diplomacy" at Compton where we all made our moves once a week, so that in the interim it was difficult to see who was talking to whom. After a year of unparalleled treachery, G emerged the winner and was extremely proud of being Diplomacy Champion of 1976. We never had quite such a good game again because several of the competitors had become somewhat inhibited after receiving phoned "Diplomacy" messages while in conversation with the director.

G was largely responsible for the design of the extremely complicated railway game "1829." "1829" has never really caught on because it takes so long to learn to play it well, and the full game takes a whole weekend to play. But it is one of the few board games where the result depends almost entirely on personal skill.

G was also a fine bridge player, capable of playing at a very high level, which only lack of frequent practise prevented him from sustaining all the time. I was privileged to be his regular partner for several years, and we had a lot of fun, as when a treasury inspector considered it an omen for the future precedence of biology over physics when Compton defeated the bridge team from the neighbouring atomic energy establishment at Harwell! However, the only time I can recall our playing a near-perfect bridge session together came sadly just a day or two before G died of a sudden heart attack in 1978. He was only 54, and several years later colleagues were still fruitfully developing the work he left in hand at that time.

Alan Dickinson's criticisms came unfortunately at the time X had commenced work with us. X came with an outstanding first-class honours degree, and he was very ambitious. However, two and a half years into his Ph.D. studies on scrapie, he was still without any positive results. The pressures on postgraduate students are considerable, and I should have been suspicious when he suddenly produced out of the blue not merely his scrapie DNA but also the spectacular changes in the synthesis of polyadenylated RNA induced in brain by scrapie. One problem was that I was not supervising X directly, his Ph.D. supervisor being Richard Kimberlin.

Richard Kimberlin is an outstanding scientist who has made many thoughtful contributions to scrapie research, particularly in terms of understanding of the mechanisms involved in the growth and spread of the scrapie agent during the incubation period before the appearance of overt disease. Thus, he has mapped the pathways by which it enters the central nervous system after a (more natural) peripheral route of inoculation, and he has discovered several chemical treatments that can modify the course of the infection and even prevent the appearance of disease in certain special cases. He also studied the passage of various strains of scrapie agent between different species of animal and discovered a Syrian hamster model with a relatively short incubation period of about 60 days. This model has been extensively used in recent years, especially by Stan Prusiner.

Kimberlin's strong personality clashed violently with that of X. They communicated mainly by means of notes left on each other's desks, and I became a sort of "piggy in the middle," trying to keep a dialogue going. In the resulting endeavour to be even-handed and fair, we gave too much rein to X. It should be added, too, that X's results were those that we would have liked to be true, and it is always easier to accept findings that are attractive than those that are otherwise. But how did X produce so many results that proved to be unrepeatable?

It may seem to the reader inexplicable that a highly skilled research worker like X could come to believe strongly in findings which subsequently proved to be totally invalid. But one possible explanation is related to the way in which scientific research is commonly reported. Gilding of the scientific lily is widely practised, and anyone who has done much research knows only too well that he never seems to be able himself to reproduce the beautiful curves and straight lines that appear in published texts and papers. In fact, scientists who would be most insulted if I accused them of cheating usually select their best results only, not the typical ones, for publication; and some slightly less rigorous in their approach will find reasons for rejecting an inconvenient result. I well remember when my colleague David Baird and I were working with a famous Nobel prize winner (Sir Hans Krebs himself) on bovine ketosis. The results from four cows were perfect, but the fifth wretched cow behaved quite differently. Sir Hans shocked David by stating that there were clearly additional factors of which we were ignorant affecting the fifth cow, and it should be removed from the analysis. We finally agreed to mention it in a footnote. Such subterfuges rarely do much harm, but it is an easy step to rejecting whole experiments or parts of experiments by convincing oneself that there were reasons that we can identify or guess at for it giving "the wrong result." In my view X fell into this latter trap, although he may never have fully realised that this was the case.

Shortly afterwards, I had to give the opening paper at a major slow virus meeting in Paris, and I had to make the sad announcement that work with our previous collaborator X must all be disregarded. But I was able to show how our subsequent work, involving Alistair Lax, Geoff Millson, and others, had repaired the damage; and our current progress in purification of the scrapie agent was at least as good as anyone's. But it was too late. The ARC ordered our programme to stop, and Alan Dickinson had the British field to himself at last. Alas, the complications involved in the setting up of a new unit have meant that little progress has been made at Edinburgh until very recently.

THE LATEST AMERICANS

Scrapie research continues to throw up interesting personalities. It is partly the nature of the beast: if you set up a large experiment and then have to wait six or eight months or even longer for disease to appear, it is not conducive to a calm life if halfway through this period you realise on reflection that there has been a fault in the experimental design. This is combined with the excitement generated by the realisation that you are working right on the frontiers of knowledge in a field important to medicine, agriculture, and basic science. You have all the ingredients for the development of any extreme quirks of temperament.

Stan Prusiner entered these troubled waters in 1972 and has since displayed a warm extrovert personality that has kept alive the scrapie tradition of generating lively discussions at scientific meetings. Personally, I have always found him friendly and helpful, but many of his opponents have resented what they consider to be his overinterpretation of his results. In particular, his introduction of the new term *prion,* to describe the fundamental scrapie unit, has been considered premature. But as we have seen in the last chapter, there does now seem to be some solid scientific basis for considering the scrapie agent as a totally new type of infective agent. The prion protein does behave differently from the normal when extracted from scrapie brain; and even if nucleic acid is directly involved in its replication, it does display other interesting properties and may be responsible for causing the lethal damage to scrapie-infected cells. Needless to say, Alan Dickinson has been critical of the prion hypothesis and has countered with his own word for scrapie—the virino.

Pat Merz, at the Mental Retardation Laboratory in New York, is a lively New Yorker, but has presented her work on the associated fibrils (SAF) in scrapie brain with restraint and in a way all can appreciate. However, her co-worker and director, Henry Wisniewski, comes rather larger than life. He succeeded a few years ago in leaving Poland to attend a conference at the same time precisely that his wife took their children out for medical treatment. Neither the Russians nor the Americans believed that this skillful escape could have been achieved without some special help, and he claimed that for years afterwards he could not go out to dinner without there being a CIA man at a table in one corner of the restaurant and a Russian agent seated in the opposite corner.

Henry Wisniewski gets through a vast amount of work with great enthusiasm, and his restless personality and nervous energy are well illustrated by his technique for watching television. On the rare occasions he finds the time, he always watches six programmes at once. Sitting in his chair with the remote channel control, he switches, at ten-second intervals, through the six programmes continuously. Probably one-sixth of their time was all that most were worth, but I found the experience at midnight after a hard day distinctly disorienting.

It will almost be disappointing when the scrapie agent is finally tamed. There has been a lot of scientific failure; the going has been tough, but, nevertheless, little by little some progress has been made. The story so far has been told by so many strikingly original individuals that it will be sad to reach the end of such a good book.

Chapter 8
Scrapie: What Does the Future Hold?

It is a commonplace that scrapie research workers expect to see the agent identified "within the next two years or so." I shall avoid falling into this trap again; but as I have pointed out earlier, it is remarkable that it still manages to hide from modern molecular biologists and it would indeed be amazing if it were not flushed out within the next 10 or 20 years.

What consequences would follow? As far as scrapie and the immediately related conditions are concerned, it should, of course, enable us to develop sophisticated diagnostic techniques that would establish the presence or absence of the disease agents in a given situation. These would be of immense value in answering questions like: "Do all sheep carry scrapie agents but only some succumb to the disease?" and, similarly, "Is Creutzfeldt–Jakob disease invariably fatal to man or do some or most recover without displaying clinical symptoms?" Fully effective disease-control measures cannot be put into operation until we know the answers to questions like these.

Another question that can be answered when we have identifiable macromolecules in our hands is the relationship between different members of the group. The epidemiological evidence (e.g., scrapie does not occur in Australia and South Africa but Creutzfeldt–Jakob disease does) suggests that scrapie and Creutzfeldt–Jakob disease probably diverged quite a long time back in evolutionary history. But this may not be so, and it is possible that sheep scrapie is a hazard for some, if not all, human individuals.

When we know the agents in molecular terms, we can approach the problem of how they enter and damage their target cells. If the avenues of initial attack (i.e., the early stages of pathogenesis) are identified, we shall be in a position to assess the hazard presented by individuals and species to each other. Is it hazardous to handle brain from a cow with BSE or to carry out a postmortem on a Creutzfeldt–Jakob patient? At present we do not really know the answers, although commonsense experience suggests that the risks involved are small for most people. If there are any risks identified, then knowledge of the molecular structures involved and of the molecular mechanisms of early pathogenesis will enable us to mount effective countermeasures. At present we can only put together tentative relationships like those shown in Fig. 9.

The advent of BSE, or "mad cow disease," has had one benefit: it has given a stimulus to research. Some of the outstanding questions will surely be answered soon as a new generation of scientists join the quest for enlightenment. Their research over the last five years has concentrated on the genetics and chemistry of the prion protein and its gene, in a variety of animals and disease situations. In general, the results have confirmed a central role for the prion protein in scrapie disease, but the overall position is still complex and unresolved. Molecular biologists, like Jim Hope in Edinburgh, should have the technical know-how to accelerate our understanding of the nature of the scrapie agent. But, as with the previous generation, they will have to learn to live with experiments of a dauntingly long-term scale.

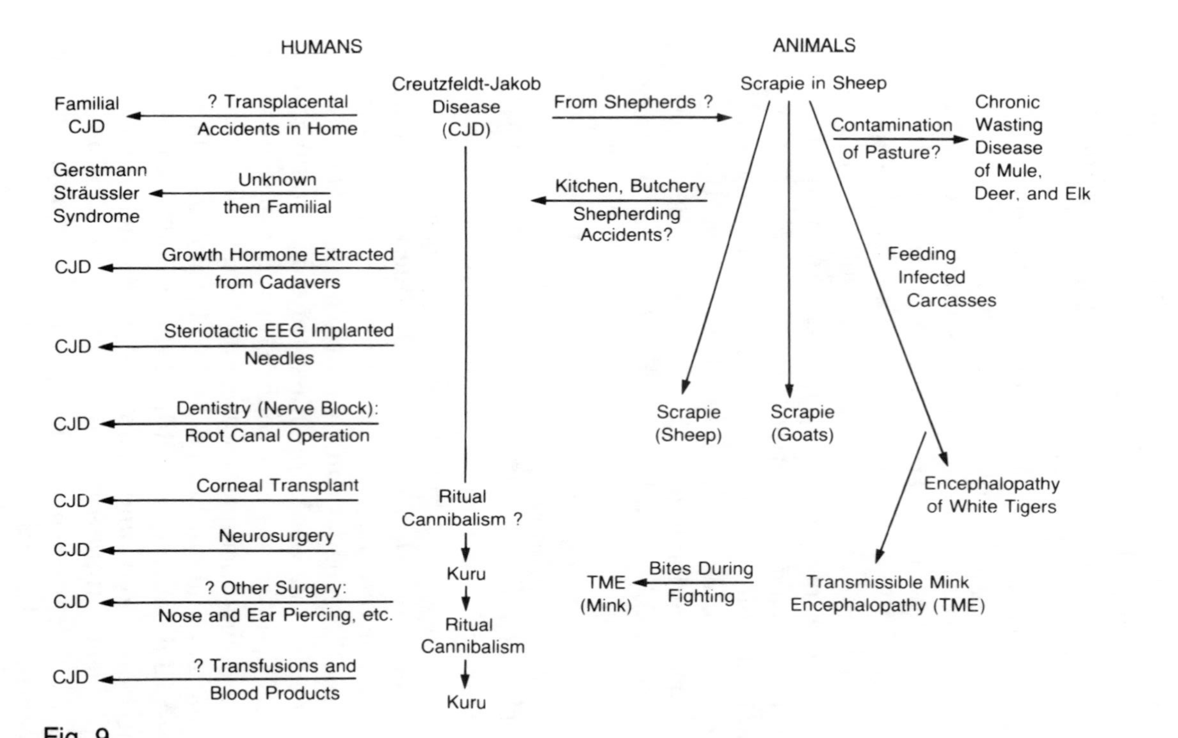

Fig. 9

Known and conjectural natural routes of transmission of the scrapie group of agents within and between species.

But we should look beyond the readily identifiable, transmissible encephalopathies. There is no doubt that the scrapie group of agents will, at the very least, present us with a new group of viruses with some unusual features in their makeup. It is most unlikely that the group we have encountered so far is exhaustive. Once we know what we are looking for, it will be much easier to find and isolate further members of the group. There are many diseases that still defy complete description: many forms of cancer, multiple sclerosis, rheumatoid arthritis, Crohn's disease, some forms of diabetes, to mention a few. In all of these, slow viruses of the scrapie group may play a part, and in complete analogy with conventional viruses and other microorganisms, there may be many nonpathogenic forms that enjoy harmless or symbiotic (mutually advantageous) relationships with large living organisms. And surely there must be the equivalent of the scrapie agent lurking somewhere in the plant kingdom.

One of the highest hopes must, however, be that better knowledge of scrapie will unlock the door to an understanding of Alzheimer's disease. When, as recently as 1907, Alois Alzheimer described the thickening and tortuosity of fibrils within the neuronal cells of a 51-year-old woman, he can have had little idea of the momentousness of his discovery. What then appeared as an isolated pathological curiosity is now known to be representative of one of the major diseases of man and one which destroys the intellect of half of us if we live to a ripe old age; it deserves much greater attention than it has yet received. The hospitals for the elderly in the U.S.A. are crammed with demented patients, and Europe is rapidly approaching a similar unenviable situation.

There is no consistent evidence that Alzheimer's disease is transmissible, but the damage observed in the degenerating brain of patients shows many analogies to the damage found in scrapie or Creutzfeldt–Jakob brains. Slow viruses may well play a part in the development of the condition, and recent research in America has shown that there is an alteration in neurofilament protein, perhaps an Alzheimer equivalent of the modification of the scrapie prion protein.

Finally, I will come clean and admit to being that unfashionable creature, a curiosity-oriented scientist. I want enlightenment above all, and desperately would like to know what the scrapie agent is. That is not to say that I want to know at any cost, but I have never believed that any advance in knowledge need be against the interests of the human race. What is the destiny of man other than a greater comprehension of the universe in which we live? Slow viruses are a part, how important a part being unknown and debatable, of that universe. Our destiny is to advance or decay, and we cannot stand still. It is true that full knowledge of the structure and behaviour of slow viruses might lead to the design of biological weapons of a truly devilish character. But that is not a path that we should choose, and the same knowledge can probably be used to alleviate several and perhaps many horrible disease conditions presently afflicting men and animals. Above all, it can give us the intellectual satisfaction of a greater understanding of, and sympathy with, a small part of this fascinating world we live in.

Further Reading

For clarity in reading, sources of information have been totally excluded from the text. The interested reader can, however, easily gain access to scrapie literature. The clinical and historical aspects of scrapie are probably best described in James Parry's excellent book *Scrapie in the Sheep* (edited by Oppenheimer and published by Academic Press Inc., 1981); the book provides a wealth of original references. The early quantitative work on scrapie in the mouse is adequately covered in my own review, "Scrapie, a prototype slow infection," published in the *Journal of Infectious Diseases* 125 (1972: pp. 427–440).

More recent work has been well reviewed by Carleton Gajdusek and Stan Prusiner, among the more recent being Gajdusek's "Chronic Dementia Caused by Small Unconventional Viruses Apparently Containing No Nucleic Acid" (*The Biological Substrates of Alzheimer's Disease*, edited by Scheibel, Wecksler and Brazier and published by Academic Press, Inc, 1986, pp. 35–54); and "The Biology and Structure of Scrapie Prions," by Michael McKinley and Stanley Prusiner, featured in vol. 27 of *International Review of Neurobiology*, published by Academic Press, Inc., in 1987.